REVISE EDEXCEL
FUNCTIONAL SKILLS LEVEL 1

English

REVISION GUIDE

Author: Julie Hughes

D1494225

A note from the publisher

In order to ensure that this resource offers high-quality support for the associated Pearson qualification, it has been through a review process by the awarding body. This process confirms that this resource fully covers the teaching and learning content of the specification or part of a specification at which it is aimed. It also confirms that it demonstrates an appropriate balance between the development of subject skills, knowledge and understanding, in addition to preparation for assessment.

Endorsement does not cover any guidance on assessment activities or processes (e.g. practice questions or advice on how to answer assessment questions), included in the resource nor does it prescribe any particular approach to the teaching or delivery of a related course.

While the publishers have made every attempt to ensure that advice on the qualification and its assessment is accurate, the official specification and associated assessment guidance materials are the only authoritative source of information and should always be referred to for definitive guidance.

Pearson examiners have not contributed to any sections in this resource relevant to examination papers for which they have responsibility.

Examiners will not use endorsed resources as a source of material for any assessment set by Pearson.

Endorsement of a resource does not mean that the resource is required to achieve this Pearson qualification, nor does it mean that it is the only suitable material available to support the qualification, and any resource lists produced by the awarding body shall include this and other appropriate resources.

Contents

A small bit of small print
Edexcel publishes Sample Assessment Material and the
Specification on its website. This is the official content and this
book should be used in conjunction with it. The questions in Now
try this have been written to help you practise every topic in the
book. Remember: the real exam questions may not look like this.

Your reading and writing tests

To do well in your Functional Skills qualification, you will need to prepare for your reading and writing tests.

Your reading test

You will have **45 minutes** in the reading test for a total of **20 marks**.

 Read all the questions
 Read the two texts
} 10 minutes

 Answer the questions (30 minutes)

 Check your answers (5 minutes)

Planning your time

Planning your time during your test is very important. Practise planning your time when you answer practice questions in this book.

In the first ten minutes of your reading test, you should:

1 **Read the questions** to work out what you need to look for when you read the texts.
2 **Skim read the texts** to find the main ideas.
3 **Underline** any information you will need in your answers.

Your writing test

You will have **45 minutes** in the writing test to answer two writing tasks worth a total of **25 marks**. One task will be worth **15 marks** and the other will be worth **10 marks**.

 Read the task and any information provided

 Produce a brief plan

 Write a detailed answer

 Check your work

Checking your work

Always read your work through when you have finished writing. You should aim to spend two minutes checking your answer is complete and correct.

When you have finished reading each task, write a brief plan for your answer, including:

• your main ideas
• notes on the audience, purpose and format.

Getting it right

Divide your time carefully between the two tasks. Leave yourself plenty of time for the second task so that you aren't rushing your answer.

Now try this

1 How many texts will you have to read in the reading test?
2 What should you do when you have finished writing your answers?

1

Reading test skills

For your reading test, you will be asked to read two texts, both about the same theme or topic. You will be tested on how well you read and understand these texts.

Reading texts

When reading the texts you should identify the following important features:

- the theme or topic
- the purpose
- the audience
- key information
- required actions.

> **text** *noun*
>
> a piece of writing that has a purpose and an audience
>
> *Examples:* letter, blog, poster, newspaper article, report, advert

Identifying important features

The heading or title will usually make the **theme** of the text clear – here it is about hairdressing.

Look for clues in the text to work out who the **audience** is. Some audiences are identified more easily than others.

Tales from the Cutting Edge

My hairdressing blog

Hi I'm Fiona and I've been a hairdresser for 10 years. I'm writing this blog for hairdressers and people who want to become hairdressers.

Cutting-edge equipment

Posted on 20th April

I was telling a customer today about my very first job in a small salon in a tiny village. Most of our customers were old ladies who wanted a cheap blow dry once a week. The equipment was cheap as the owner needed to keep her prices as low as possible. The problem was, we were constantly replacing the equipment as it was poor quality. At Kids' Cutters we use top-quality hairdryers, scissors and shampoos. Our clients are often as

Sign up to my newsletters!

Archive:
- January
- February
- March
- November
- December

Search the text for command words to work out if any **action** is required. This text asks you to sign up to a newsletter.

The title of this text also tells you that is is a blog. Blogs can have several different **purposes**, but, like this blog, they usually aim to inform.

Further reading will reveal **key information** and details. For instance, here the first entry is about hairdressing equipment.

In your reading test, you can make notes on the texts to help you answer the questions but these notes won't be marked.

Now try this

1 What can a heading or title help you to identify?
2 Can a text be suitable for more than one audience?
3 Name **two** things used to draw attention to important information in a text.

Multiple choice questions

The reading test contains a mixture of multiple choice and short response questions.
These questions will test whether you have read and understood the two texts in detail.

Answering multiple choice questions

For multiple choice questions, you will be asked to select **one** or **two** correct answers from a choice of **four** or **five** options.

Use clues in the question to help you work out how many answers are required. Put a cross in the box (or boxes) next to the correct answer(s).

> For the online test, you should click the box next to the correct answer. You can change your answer during the text by clicking in another box.

Worked example

1 The **main** purpose of Text A is:

to inform the magazine about the Academy's survey about positive role models ☐

to tell students not to read magazines ☒

to inform students about the use of positive role models ☐

to persuade the magazine to start using positive role models ☒

(1 mark)

> The first question about each text will ask you to identify the **main** idea, or purpose, of the text. You should only choose **one** answer for this question.
>
> If you change your mind about the answer, don't worry.
> 1 Put a line through the wrong cross.
> 2 Mark your new answer with a cross.

Worked example

2 Identify **two** features from the list below that show that Text I is a web page.

Menu tabs ☒

Images ☐

Punctuation ☐

Website address ☒

Title ☐

(2 marks)

> **Feature**
> a distinctive characteristic of something.
> *Similar words:* quality, property, attribute

> Here, the question asks for **two** features, so **two** crosses have been put in the boxes next to the two correct features.

> There are 2 marks available in total, one for each correct answer.

Now try this

1 How many options could you be given in a multiple choice question?
2 What should you do if you want to change your answer?

Short response questions

For your reading test, you will also need to write answers to short response questions to show that you have read and understood the text.

Answering the question

For short response questions you should:

- read the question carefully and identify any key words
- work out how many answers you need to write
- skim the text, looking for key words that are used in the question
- gather information from the text for your answer
- write your answer focusing only on the relevant information and what is needed.

Your answer should be based on the information given in the text only.

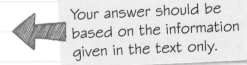
Go to page 33 to revise using information in an answer.

Worked example

1 According to Text B, give <u>one</u> reason why a <u>driving licence</u> is necessary.

You do **not** need to write in sentences.

to collect supplies from the warehouse

(1 mark)

Here, the student has underline the key words **driving license** and focused their reading to reveal the correct answer.

There is 1 mark available so the student has written one answer.

Worked example

2 Your friend is thinking about applying for the job of Window Display Assistant but is worried that he does not have enough <u>experience</u>.

Using Text B, give <u>**two**</u> types of <u>training</u> that are given to Window Display Assistants to support them when they start work.

You do **not** need to write in sentences.

1 window display training
2 sewing-machine training

(2 marks)

The key words here are **experience** and **training**. The student skims Text B to find where the word training, or a similar word, appears.

You could be asked to give **two** answers to a short response question.

You should write down only the key information that is needed for the answer. Don't waste time writing complete sentences.

Bullet points are always a good place to start looking for key information in the text.

Now try this

1 Write down **three** rules for answering short response questions.
2 Where should you find the information to answer the question?

Reading the question

In the test, you should read each question carefully to make sure you know exactly what it is asking you to do.

Reading test questions

When reading the question, you should identify the following things to help you with your answer:

- the type of question
- key words — in the reading test, important words are sometimes emphasised in bold
- the number of answers needed.

> All your answers must use information from the text. Even if you know something about the topic, you will not get marks if you use information that is not in the texts provided.

Worked example

This is a multiple choice question so you would need to put a cross next to the correct answer.

Here you are asked to find the **main** purpose. The student must identify the most important purpose.

Read all the options carefully. Some multiple choice options start with the same words or phrases. Both A and C start with 'describe'.

There is only one answer needed for this question.

1 The **main** purpose of Text B to:

A   describe what is needed to become a Window Display Assistant   ☒

B   persuade people to become a Window Display Assistant   ☐

C   describe the criteria to become a sixth form student   ☐

D   tell students not to work at Fashion Fun   ☐

(1 mark)

Worked example

This is a short response question so you would need to write a short answer.

Don't waste time writing full sentences. Practise making your answers as short as possible.

Here, the question asks you to list two features. There is 1 mark for each correct feature.

2 Paragraphs are one other feature used in Text B. List **two** other features of Text B that help to make it easy to understand.
You do **not** need to write in sentences.

1   Bullet points

2   Table

(2 marks)

Now try this

1 How many answers should you give for a 2-mark question?
2 Why should you use **only** information provided in the text to answer questions?

Skimming for details

Skim reading a text helps you to locate key information and identify the main ideas quickly. When you skim read a text you should focus on key parts, such as:

- titles or headings
- the first sentence of each paragraph
- bold, underlined or highlighted text
- images
- numbers and bullet points.

Getting it right

In the test, you should read the whole text carefully, not just the title or heading.

Worked example

Read Text A and answer the question below.
The **main** purpose of Text A is:

A to persuade the magazine to use positive role models ☐

B to inform the magazine about the latest fashions ☐

C to describe what a positive role model is ☐

D to instruct the magazine to use less positive role models. ☐

Skim reading tips

Skim reading is an important skill that you should learn and practise before your reading test.

To skim read effectively you need to:

- Read the questions first so you have an idea of what you are looking for.
- Focus your reading on key parts of the text. You don't need to read every word when you are skim reading.

Text A

Dear Editor,

Positive role models in the media

I have decided to write to you to complain about the lack of positive role models in your magazine.

Your magazine is for people between the ages of 13 and 18, who often feel

under pressure to 'follow the crowd' so they can fit in with their friends.

They look to your magazine for guidance. Your glamorous stories and photographs seem to be all about appearance, even though there is much more to life! You should be inspiring young people to be happy, healthy and ambitious.

The title, or heading, can give you an idea of what the text is about. Remember to read on further to understand the text fully.

The first sentence of this text tells you that the writer is writing to complain about something. If you read on, the writer describes their complaint and tries to persuade the editor of the magazine to address the problem.

Now try this

1 List **two** key places to skim for information in a text.
2 Why is it a good idea to read the questions before skimming a text?

Underlining

In your reading test, underlining and circling key words and phrases in the questions and texts can help you to access important information quickly and easily.

Underlining important information

When reading the text, you should underline important information, such as:

- key words in the text and the question
- numbers
- instructions
- words relating to the purpose of the text.

Getting it right

Before you start underlining the text, you should do the following so that you can focus your reading:

✓ read the questions

✓ pick out the key words in each question so that you know what to look out for in the text

✓ skim read the text to find the main ideas.

Worked example

Read the extract from Text E (page 78) on the right.

Identify **two** things the Estrick Healthy Living Advice Centre does to help residents <u>achieve</u> a <u>healthy lifestyle</u>.

Set up a free voucher scheme

Provided free health checks

✓ The student has underlined the key words 'healthy lifestyle' and 'achieved'.

✓ The student has written two correct answers.

You might find more correct answers in the text than you need. Don't waste valuable time giving more answers than you have been asked for.

Text E

The Estrick Healthy Living Advice Centre was set up five years ago by the local council, and we are based in the Town Hall. Our aim is to get local residents following a <u>healthy lifestyle</u> by eating well and exercising more.

What we have <u>achieved</u> so far:

- <u>Set up a free voucher scheme for access to four local gyms</u>
- <u>Provided free health checks for anyone who is worried about their weight</u>
- <u>Provided outdoor gym equipment at the local park</u>
- <u>Set up a healthy living website that offers tips on how to eat healthily</u>

Go to page 78 to read the whole of Text E.

Now try this

Read another part of Text E on the right and the question below. Underline or circle any words in the extract that you think you could use to answer the question below.

List **two** ways in which volunteers can help the Healthy Living Advice Centre. You do **not** need to write in sentences.

Text E

How you can get involved

We need volunteers to help us spread the word about healthy lifestyles. Volunteers can help in many ways: by raising funds, by working in our local centre, by testing the recipes on our website or by helping us to plan guided walks. You could also help by delivering leaflets – that way you will also help yourself by keeping fit!

Online tools 1

If you are completing the Functional Skills tests online, you will need to understand how the test works.

Before you start your online test:

- Read the instructions about how to use the test.
- Make sure you know what all the icons do.
- Make sure you can read the test clearly and easily.

Useful icons

You can click this Time icon to find out how much time you have left in your test. The time will appear in the bottom left-hand corner.

The timer does not stop when you click on the Help icon. Make sure you know how to use the test before you start. You will be reminded when you have 15 minutes left, and again when you have 5 minutes left in the test.

Time Help Review Flag Previous Next Exhibit Translation Quit Status

You can click this Help icon if you want a demonstration of how the online test buttons work.

The Previous and Next icons move you from question to question.

Be very careful with the Quit icon. If you click on it and then select 'yes', you will not be able to return to the test even if you haven't finished!

Changing the test settings

Click the ✚ button in the bottom left-hand corner of the screen to open the Settings box.

Use the colour reset button and the zoom reset button to go back to the original test screen.

Click the arrows to move around the page when you are zoomed in.

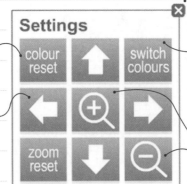

Click this button to close the Settings box.

Click the switch colours button to change the colour of the test to make it easier to read.

Click the magnifying glass icons to zoom in and out of the test.

Now try this

1 How do you find out how much time is left?
2 What can you do if you can't read the test clearly?

Online tools 2

There are useful tools on the online test that can help you plan your answers and your time.

The notepad tool

You can make notes to help you to remember key information and plan your answer.
The notepad tool takes you to a screen where you can:

- underline questions
- underline words and phrases in the reading texts
- make a plan for answering your writing tasks.

Click on this icon to open the notepad. Anything you type in this box will **not** be marked.

This button will let you underline key words in the texts or questions. Make sure the shaded box covers the text you want to underline.

Click one of these buttons to choose a colour for your notes. Choose a colour that is different from the text, so that you can see it clearly when writing your answers.

This button clears everything in the notepad and all your underlining. Make sure you don't press Clear All before you have finished and checked your work.

Formatting text

In the writing test, you can change the font and the presentation of text for emphasis.

Go to pages 38 and 39 to read more about writing test tasks and skills.

Flagging tricky questions

 If you are unsure how to answer a question, **flag it** and move on to the next question. Click this Flag icon so that you can come back to the question later.

Before you check your answers, go back to any questions you have **flagged**. You can go back to questions you have flagged by clicking the **Review** button.

Now try this

1 What **three** things can you use the notepad tool for?
2 What should you do if you are struggling to answer a question?

Putting it into practice

You now know what to expect from the questions in the reading test. Prepare for your test by practising the following:

- reading and understanding questions
- skim reading texts
- underlining key information while you read

- answering multiple choice questions
- answering short response questions.

Read the text extract from Text F and the test-style questions and look at how a student has prepared for them. The full version of Text F is on page 79.

Text F

What's there

The nature reserve opened three years ago and has since become the most visited tourist site in the Estrick area, with over 5,000 visitors per month. The money raised from renting the holiday properties helps to improve the habitats of many protected bird species.

During my stay at the reserve I saw reed warblers and red kites. They are only two of the protected species that have come back to the area because of the conservation work done at the reserve. Sharp-eyed birdwatchers can look out for marsh harriers, lapwings and teals, and may even be lucky enough to see the reserve's bean geese.

Visitors using the cottages will be encouraged to take part in the day-to-day running of the reserve by helping to plant trees.

Worked example

1 According to Text F, what has the reserve become since it opened three years ago?

The most visited tourist site in the Estrick area

✓ The student has noticed that **three years ago** is a key phrase in the question and found it in Text F.

✓ The answer includes the right amount of detail.

Worked example

2 Identify **two** birds that the writer saw at the reserve.

A reed warblers ☐

B marsh harriers ☒

C lapwings ☐

D red kites ☐

E teals ☐

✗ The question asks for two birds, and the student has marked only one answer.

✗ Here, the student has looked for types of bird without reading the question carefully first. The question asks for birds that the writer saw, **not** birds that visitors might see.

Now try this

Read the whole of Text F on page 79 and answer question 2 above.

Read the questions carefully before reading the text. Then read them again before writing your answer. This way you will have two chances to get your answer right!

Understanding the main idea

You will need to identify and understand the main ideas of the texts in the test so that you can answer all the questions.

Understanding texts

You can work out the audience, purpose and what type of text you are reading by understanding the main idea of a text. Look at the extract from Text B (page 75) and the worked example below to see how you can relate the main idea to the type of text, audience and purpose.

Text B

Person Specification
<u>Window Display Assistant</u> for all Estrick branches of Fashion Fun.

	Criteria	How identified
Qualifications	• Level 2 Functional Skills Maths and English. • At least two GCSEs, including a Design Technology subject such as Textiles, Graphic Design or Resistant Materials.	Certificate

Worked example

The **main** purpose of Text B is:

A to describe Fashion Fun ☐

B to inform you about all Estrick branches of Fashion Fun ☐

C to describe what skills are required to become a Window Display Assistant ☒

D to instruct you to produce an attractive window display. ☐

The student has read the heading and identified that Text B is a person specification.

As with the majority of specifications, the purpose of Text B is to describe. Text B is presented in a table so that the main points (the descriptions of the criteria) are easy to find.

Further reading shows that the specification is for people who want to become a Window Display Assistant.

Turn to pages 15–18 to revise identifying the purposes of texts.

Now try this

1 How can you work out what type of text you are reading?
2 List **two** of the main text purposes you might come across in your test.

Had a go ☐ Nearly there ☐ Nailed it! ☐

Identifying the main idea

In your reading test you will need to identify the main points in a text.

Reading the title

A title or heading can give you a clue as to what a text is about, but you need to read the whole text carefully to understand its main idea.

1 Sometimes, a title or heading suggests a clear main idea, such as the title on the right from Text C on page 76.
Reading the rest of Text C reveals more specific details about the main idea that is suggested in the title - how to make your home safe.

2 Sometimes, the title or heading is less clear, such as the heading on the right from Text A on page 74.
Reading the whole of Text A reveals how the writer feels about positive role models in the media and what they want the reader to know or do.

To identify and fully understand the main idea in the test, you need to read the whole text carefully, not just the title.

SAFETY IN THE HOME

Positive role models in the media

Finding the main idea

To help you work out the main idea, you need to read the whole text carefully and look for key words and phrases, images and other important features.

Look at the extract on the right from Text F on page 79. The word '**getaway**' in the title suggests that the main idea is holidays. By reading the first sentence it becomes clear that the main idea is more specific — holidays at Estrick Nature Reserve.

Text F

Thinking of a natural getaway?

By John Smith – Travel writer

John Smith spent time at one of the three holiday properties now available at Estrick Nature Reserve – an area of outstanding natural beauty within walking distance of stunning coastline walks.

Look at Text H on page 81.

1 What is the main idea in the text?
2 What makes you think this is the main idea?

Identifying types of text

In your reading test, you will need to read and understand two different types of text. This page will help you to recognise:

- letters and emails
- articles
- websites
- blogs.

Letters and emails

Letters and **emails** can be formal or informal. Look at the table below to see some of their similar features.

	Letters	Emails
Address	have an address, e.g. 36 Penny Lane, Leeds, LS10 1EA	have 'to' and 'from' boxes containing email addresses, e.g. a.bell@example.com
Subject	appears after the address	appears after the 'to' and 'from' boxes
Greeting	Dear...	Dear... (formal) Hi... (informal)
Sign-off	Yours sincerely (formal) Thanks (informal)	Yours sincerely (formal) Thanks (informal)

Websites

Websites have different features to make it easy for you to find information.

Navigation bar – shows the website's address.

Hyperlinks – clicking on a hyperlink takes you to another web page or website.

Search box – lets you search for information that might be on the website.

Blogs

Blogs are similar to websites, but information is presented in **dated sections**. They are usually written in an informal or conversational style and they feature links after each entry so that readers can respond.

Articles

Articles are usually found in newspapers and magazines. Look at the extract from Text F (page 79) below to see the typical features of an article.

Text F

Headline gives the main idea.

The introduction is in **bold** to make it stand out.

Sub-headings point out the main sections.

Thinking of a **natural** getaway?

By John Smith – Travel writer

John Smith spent time at one of the three holiday properties now available at Estrick Nature Reserve – an area of outstanding natural beauty within walking distance of stunning coastline walks.

What's there

The nature reserve opened three years ago and has since become the most visited tourist site in the Estrick area, with over 5,000 visitors per month. The money raised from renting the holiday properties helps to improve the

Where to stay

The holiday accommodation sits in five acres of stunning landscaped grounds with its own peaceful lake, tennis courts and indoor swimming pool.

Whilst there, I enjoyed a tasty but healthy lunch snack at the clubhouse.

Pictures catch the reader's attention.

Columns are used to break up large amounts of text.

Now try this

1 Look at Text I on page 82. Paragraphs are one type of feature that is used in Text I. Find another **two** features that tell you what type of text it is.

2 Make a checklist of the features usually found in letters, emails, articles, websites and blogs.

3 Practise identifying different types of text using your checklist.

More types of text

In your reading test, you will need to read and understand two different types of text. This page will help you to recognise:

- leaflets and posters
- reports and factsheets
- specifications.

Leaflets and posters

Leaflets and **posters** can have a variety of different purposes, such as to inform or to persuade. The main ideas of a leaflet and poster are usually presented clearly and in an eye-catching way for the reader.

Look at Text E on page 78 to see an example of a leaflet. Text E contains the typical features you would expect to see in a leaflet, such as:

- headings and sub-headings
- images
- bullet points
- columns.

> Leaflets often use descriptive language to make the reader feel a certain way. Text E features words, such as 'tasty' and 'fun' to make the main ideas of the text appeal to the reader.

Reports and fact sheets

Reports and **fact sheets** present information in a formal way. They use certain features to make key information clear and easy to find, such as:

- facts
- statistics
- charts
- bullet points
- tables.

> Fact sheets and reports present facts and information as evidence to inform, instruct or persuade. Text J on page 83 is a report. It features lots of facts, such as 'Donations have increased by 50 per cent'.

Specifications

A **specification** is a written description of requirements for a particular thing, such as a job or a product design. Specifications are often presented in a table to help the reader find information quickly and easily.

> Text B on page 75 is a specification for a job. The requirements are listed in bullet points in a table so that they are easy to read. Look out for key words such as 'essential' and 'advantage'.

Now try this

1 Which features can be used to make a leaflet or poster look exciting?
2 List **two** features of a report.
3 Why are specifications usually presented in a table?

Texts that inform

For your reading test, you need to be able to recognise the purpose of a text. The purpose of some texts is **to inform**. Texts that inform the reader use factual information to tell the reader something.

Identifying the texts that inform

Look out for informative writing in leaflets, fact sheets and reports. These types of text are used to give the reader important information about something. Read the extract from Text H (page 81) below and look at what makes this fact sheet an informative text.

Text H

Restaurant Work Experience Fact Sheet

You will need to wear:

Your chef's jacket

Sensible shoes or boots

You should bring:

- signed parental consent form
- completed health questionnaire
- £5 to cover food costs
- white shirt and black trousers
- notebook and pen.

For the third year running, Estrick College's catering students have been offered the chance to spend two weeks working at Estrick Hotel. This is a fantastic opportunity to get further knowledge about the catering industry and gain hands-on experience at a top-class hotel kitchen.

Built in 1900, Estrick Hotel has been serving delicious evening meals from its Conservatory Restaurant for over 100 years. Last year, the hotel added a new extension with a café serving light lunches and afternoon teas. The hotel also has a snack bar in the grounds, which provides refreshments for users of its award winning, 18-hole golf course.

Language

Texts that inform usually use **formal language** rather than slang. For example, the words 'opportunity' and 'refreshments' give Text H a formal tone.

Informative texts usually contain a lot of facts - information that can be proved to be true. Text H contains facts throughout, such as 'Built in 1900' and 'award-winning golf course'. These facts tell the reader that Estrick Hotel is a good place to do work experience.

Layout

Texts that inform often contain graphs, tables and bullet points that present lots of information in a way that is easy to read. Here, the fact sheet uses **bullet points** that make it easy to find information. The image of the chef with labels provides the reader with information in a way that is clear and easy to understand.

Now try this

1 Look at the rest of Text H on page 81.
   (a) Find **two more** facts that suggest its main purpose is to inform.
   (b) Find **two more** examples of formal language that suggest its main purpose is to inform.
2 What is Text H informing the reader about?

Texts that instruct

For your reading test, you need to be able to recognise the purpose of a text. The purpose of some texts is **to instruct**. Texts that instruct tell the reader how to do something.

Type of text

Instruction texts can take many different forms, including recipes, manuals and directions.

- Recipes **instruct** the reader how to prepare a meal.
- Manuals **instruct** the reader how to make or use something.
- Directions **instruct** the reader how to get from A to B.

Language

You can recognise an instruction text by the following language clues:

- clear language that readers will be familiar with
- concise sentences that are quick to read and easy to follow
- command verbs telling you what to do, e.g. 'make', 'turn' and 'put'.

If you are doing the online test, you could be asked to highlight language that relates to the text's purpose.

Layout

 Tables, bullet points and numbered lists are often used for instruction texts to show the information in the **correct order**.

 Pictures and diagrams are sometimes used alongside written instructions to clarify what is meant.

 Instruction texts are sometimes presented in columns so that more information can be included.

Text D

These instructions are so easy - you can set up your farm all on your own!

1. Fill a large plastic bowl with water. Be careful not to use hot water as this might make your worms grow too quickly.

2. Put the worm mixture into the water. Don't worry if it sinks to the bottom in one big lump – it will soon start to look like worms!

3. Follow the pictures on the enclosed instruction leaflet to put the farm together.

4. After 24 hours (one whole day and night) remove the worms from the water. Put them on pieces of kitchen paper and leave them to dry for an hour.

Put the worms into the farm and watch them grow!

Go to page 77 to see the whole of Text D.

Now try this

1. Look at Text C on page 76. Find **two** features that tell you that the purpose of the text is to instruct.
2. Explain why Text C uses these two features.
3. Make a list of command verbs used in the text.

Texts that describe

To understand a text you need to work out its main purpose. Some texts are written to describe. They tell readers what something is like.

Type of text

Articles, leaflets, advertisements and blogs are examples of descriptive texts. These types of text give information, like an informative text, but they are often more entertaining.

Text F

Thinking of a **natural** getaway?

By John Smith – Travel writer

John Smith spent time at one of the three holiday properties now available at Estrick Nature Reserve – an area of outstanding natural beauty within walking distance of stunning coastline walks.

What's there

The nature reserve opened three years ago and has since become the most visited tourist site in the Estrick area, with over 5,000 visitors per month. The money raised from renting the holiday properties helps to improve the habitats of many protected bird species.

Where to stay

The holiday accommodation sits in five acres of <u>stunning</u> landscaped grounds with its own <u>peaceful</u> lake, tennis courts and indoor swimming pool.

Whilst there, I enjoyed <u>a tasty</u> but healthy lunch snack at the clubhouse. As this sits on the banks of the lake, I didn't need to miss a minute of the breathtaking views.

Language

Descriptive writing uses lots of **describing words and phrases**. Words that describe things such as size, colour and shape help **create a picture** for the reader. Words that describe can make readers **feel** a certain way.

Layout

Most descriptive texts are presented in paragraphs, rather than bullet points or tables. This is so that it can build up a picture for the reader.

Pictures are often used to make the text eye-catching and to break up the text.

Now try this

1 Look at the rest of Text F on page 79. Find **two** more examples of layout and language that suggest it is a text to describe.
2 List **three** describing words you can find in the text.

Texts that persuade

Some texts are written to **persuade**. The writer will use reasons and make clear points in the text to try to get readers to believe something or to do something.

Type of text

Advertisements, leaflets and posters use persuasive writing. These types of text may want to persuade readers to buy things, to do something or to believe something.

Look out for persuasive writing in letters and newspaper articles too.

Getting it right

To work out why a text has been written, think about:
- the type of text
- how it is presented
- the type of language used.

In your test, the questions about purpose might use words with a similar meaning to persuade. Look for words such as **encourage** or **advertise** or **ask**.

How text is presented

Persuasive writing can be found in many styles of text. Persuasive writing often contains lots of details to make readers feel a certain way. Each new point the writer makes should have its own paragraph.

Type of text language

Persuasive writing often contains **facts and statistics**. These help to make the content believable and the writer seems like an expert.

Quotations are used to add importance to writing. They help support the points being made and make them believable.

Sometimes the text will ask the reader questions. This encourages the reader to think deeply about the topic.

Text A

Your magazine is for people between the ages of 13 and 18, who often feel under pressure to 'follow the crowd' so they can fit in with their friends. They look to your magazine for guidance. Your glamorous stories and photographs seem to be all about appearance, even though there is much more to life! You should be inspiring young people to be happy, healthy and ambitious.

At our Academy, we are very disappointed by the lack of positive role models for young people in the media. We carried out a survey and found that over 80 per cent of our students felt under pressure to look or act a certain way because of what they read in magazines like yours. One student said, 'I can't relate to models and film stars. Why is there never anything for sporty people?'.

We think that you should use models with varied appearances, backgrounds, and interests so that there is something for everyone.

You would benefit from making this change by gaining new readers. It would also encourage more parents to buy your magazine for their children. But most of all, think about the positive messages you could be sending out to young people.

Go to page 74 to read the rest of Text A.

Now try this

Look at Text I on page 82. Find **three** examples of facts and details that suggest it is a text to persuade.

Putting it into practice

In this section, you have revised:

- understanding the main idea
- identifying the main idea

- identifying the text type
- identifying the text purpose.

Look at the test-style question below and read a student's answer.

Worked example

The **main** purpose of Text J is:

A <u>to inform</u> readers how the Animal Shelter has spent its funds [✱]

B to ask readers to volunteer at the Animal Shelter [ ]

C to tell young people to donate [ ]

D <u>to inform</u> the reader how the Animal Shelter has raised its funds. [✗]

(1 mark)

✓ The student did not read the question carefully to begin with, but noticed and corrected their error. A and D look very similar, but D is the correct answer.

Look at pages 6 and 7 for a reminder on reading and annotating skills.

Text J

Dear Charity Fundraisers,

I am writing to give you the details of our fundraising, from October 2014 to October 2015. Together, we <u>raised</u> a lot of money for animal welfare projects in 2015, but we know that you   *facts and statistics = persuasive?*   can help us raise even more money in 2016.

The chart on the right shows the amount of money raised by each fundraising activity. Read on to find out more.

Donations

The animal shelter still relies mainly on donations from the public. Donations have increased by 50 per cent and have now reached over £10,000 since we added the new online donation feature on our website. This means that people do not have to leave home to donate. It has also encouraged younger people to donate. Online donations are higher than those received in collecting boxes.

✓ The student has found examples of informative writing, such as facts and statistics.

✓ The student has noticed that the key word 'raise' appears in the text and the question.

Getting it right

Always read every question carefully

- Some of the answer options will be similar.
- Some answer options might seem to be the right one because there may be small examples of this type of writing.
- Always find the main ideas first – this will help to you identify the main purpose.

Now try this

Identify and highlight the main ideas in the text above. This will help you to understand why D is the correct answer.

Understanding presentation

Understanding why text is presented in a certain way is key to identifying the type and purpose of a text. This page will help you to identify and understand presentation features.

presentation *noun*
the style in which something is displayed
Similar words: appearance, arrangement, organisation, display, layout

Identifying presentation features

You need to be able to correctly identify and name presentation features in your reading test. Look out for the following presentation features.

logo

columns

bullet points

The Estrick
Healthy
Living Centre

Estrick Healthy Living Advice Centre

was set up five years ago by the local council, and we are based in the Town Hall. Our aim is to get local residents following a healthy lifestyle by eating well and exercising more.

What we have achieved so far:

► Set up a free voucher scheme for access to four local gyms

► Provided free health checks for anyone who is worried about their weight

► Provided outdoor gym equipment at the local park

► Set up a healthy living website that offers tips on how to eat healthily.

What we are planning next:

► A series of guided health walks

► A fun day for families

Eat healthily

Start following a healthy lifestyle straight away by trying these menu options:

Vegetarian options

► Tofu hummus with carrot sticks

► Smoky aubergine tagine with lemon

► Pears baked with honey

Low fat option

► Grilled mushrooms

► Thai red duck with sticky sesame rice

► Fruit salad with frozen strawberry yoghurt

subheadings

images

The purpose of features

Writers use presentation features to:

• grab your attention

• tell you what a text is about

• separate information into sections

• help you understand the text

• focus your attention on certain words and phrases.

In the reading test, look carefully for any of the presentation features above. You should also look out for:

- numbered lists

- tables and charts

- text size, font and colour

- text boxes.

Now try this

List **two** presentation features that are used in Text G on page 80.
You do not need to write in sentences.

Paragraphs, columns and fonts

This page will help you to understand why writers use different fonts, paragraphs or columns.

Understanding text formatting

If certain words or phrases in a text look different from the rest, the writer wants them to stand out to the reader. Formatting also makes text look more interesting and draws readers in. Look out for the following types of text formatting to help you spot key words and ideas.

- different or embellished fonts

- variations in text size and colour

- bold.

The font looks as if it has been stamped on to the text. This makes it look official and shows that the text should be taken seriously.

Paragraphs and columns

Paragraphs split writing into shorter sections. A new paragraph is started when a new topic or idea begins.

The first sentence in a new paragraph introduces a new idea.

Here, a new paragraph is started to tell the reader where to stay – it is emphasised by a bold sub-heading.

Text F

John Smith spent time at one of the three holiday properties now available at Estrick Nature Reserve – an area of outstanding natural beauty within walking distance of stunning coastline walks.

What's there

The nature reserve opened three years ago and has since become the most visited tourist site in the Estrick area, with over 5,000 visitors per month. The money raised from renting the holiday properties helps to improve the habitats of many protected bird species.

During my stay at the reserve I saw reed warblers and red kites. They are only two of the protected species that have come back to the area because of the conservation work done at the reserve. Sharp-eyed birdwatchers can look out for marsh harriers, lapwings and teals, and may even be lucky enough to see the reserve's bean geese.

Where to stay

The holiday accommodation sits in five acres of stunning landscaped grounds with its own peaceful lake, tennis courts and indoor swimming pool.

Whilst there, I enjoyed a tasty but healthy lunch snack at the clubhouse. As this sits on the banks of the lake, I didn't need to miss a minute of the breathtaking views.

Woodpecker Cottage sleeps two in a beautiful, king-size four-poster bed. It is a pretty, little property, cosy and peaceful, making it ideal for a romantic break for two.

I was lucky enough to stay in the largest and most luxurious property offered by Estrick Nature Reserve.

The writer has split the text into two **columns** to break the text up so that it is easier to read and so that more information can fit on the page.

New paragraphs don't always need sub-headings. The descriptions of the two cottages have been separated into two paragraphs here.

Now try this

1 How can text be formatted differently to make it stand out more?

2 Find **two other** texts from pages 74 to 83 that split text into columns.

Titles, headings and lists

This page will help you to identify titles, headings, lists and bullet points and understand why they are used.

Titles and headings

Titles and **headings** attract a reader's attention and show what the text is about.

Sub-headings are used to break up smaller sections of the text. Sub-headings help readers to find information they need quickly and easily.

> Titles, headings and sub-headings give you a brief idea of what the text is about, but you should read the whole text carefully so that you don't miss any key information!

Numbered lists and bullet point

Numbered lists and **bullet points** split information into small pieces. This makes the text clear and easy to read.

Numbered lists are often used for instructions as it is important that the text is read in a particular order.

Text I

ESTRICK VOCATIONAL COLLEGE

Building trades department

| Home | Courses | Careers | Term Dates | Application |

Estrick Vocational College was set up 20 years ago to train skilled people to work in local businesses. The Building Trades Department was started 5 years ago and now it runs over 20 courses, including both full-time and part-time.

We have invested in new facilities. Last year, a new building trades centre was added. It has workshops for bricklaying, joinery, painting and decorating, plastering, plumbing and electrical installation work. We aim to prepare students for life on a real construction site. We also built new classrooms and an assembly hall. This year, we have spent over £2 million on a state-of-the-art computer suite. These improvements will support students with the academic side of their college courses.

Why choose Estrick College?
- All our courses are industry-approved.
- Our workshop facilities meet all health and safety standards.
- We provide all materials and tools.
- Our staff are fully qualified and friendly.
- We offer support with English, Maths and IT skills.
- Over 75% of students find a construction industry job within one month of leaving college.

Why do employers choose Estrick College students?

Our students are fully prepared for the workplace. All students have experience of working on real construction sites. For the past 3 years all our students have left the college with industry-approved qualifications in their particular trade subject.

This website uses the name of the college as a main heading.

The sub-headings here introduce particular bits of information.

The bulleted list separates important information about Estrick College into brief points tha stand out.

Notice the two different sub-headings use different forms of presentation to draw the reader's attention to them.

Now try this

1 Why is it important to read the whole text and not just the title or heading?
2 Find **two other** texts from pages 74 to 83 that use numbered lists or bullet points.

Tables and charts

Tables, **graphs** and **charts** are used to make information in a text easier to understand.

Tables

Some texts show information in a table. Tables are mostly used for factual texts. They make information easy to find.

In this text, if you read the headings of the columns and then the rows, you can easily find out if you have the qualifications and experience to apply for the job.

In the table below, from Text B on page 75, the first column and row contain the most important information – here you can quickly see that there is no point in reading on if you do not have at least a Level 2 Functional Skills Maths qualification.

Person Specification
Window Display Assistant for all Estrick branches of Fashion Fun.

	Criteria	**How identified**
Qualifications	• Level 2 Functional Skills Maths and English. • At least two GCSEs, including a Design Technology subject such as Textiles, Graphic Design or Resistant Materials.	Certificate
Relevant experience	• No previous experience with window displays is required as training will be provided. • Previous retail experience would be an advantage.	Interview

Rows ———

Columns ———

Graphs and charts

Graphs and charts present information in a visual way. Like tables, they make information easy to find and understand.

Graphs and charts usually have a heading explaining what they are showing. Look at this first to help you to understand the point of the graph or chart.

The chart on the right is taken from Text J on page 83. The chart makes it easy to see which fund-raising method is the most successful. The higher the bar, the more money was raised.

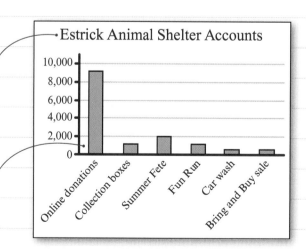

Estrick Animal Shelter Accounts

(bar chart: Online donations ~9,000; Collection boxes ~1,000; Summer Fete ~2,000; Fun Run ~1,000; Car wash ~500; Bring and Buy sale ~500)

List **two** reasons why tables, charts and graphs are used to present information.

Images and graphics

Images and graphics are used to make texts more interesting and appealing. Look at how images and graphics are used in Text E below.

Graphics

Graphics can be images, logos, charts or diagrams. Graphics are used to give extra information, or to make a text look more interesting. They often have a caption to tell you something about them.

Images

Images, or pictures, are used to make a text more appealing. They also give more information about the topic of the text.

In Text E, the image is used to show how attractive healthy meals can look.

Text E

The Estrick
Healthy
Living Centre

Estrick Healthy Living Advice Centre

was set up five years ago by the local council, and we are based in the Town Hall. Our aim is to get local residents following a healthy lifestyle by eating well and exercising more.

What we have achieved so far:

▶ Set up a free voucher scheme for access to four local gyms

Eat healthily

Start following a healthy lifestyle straight away by trying these menu options:

Vegetarian options

▶ Tofu hummus with carrot sticks
▶ Smoky aubergine tagine with lemon
▶ Pears baked with honey

Low fat option

Getting it right

In your reading test, you should comment on why or how a presentation feature, such as an image or a photo, is being used. You won't get any marks for just writing 'tofu hummus'!

Logos

Logos are used to make it easy for the reader to identify who has produced the text. Logos often represent the main aims and ideas of a company. Here, the three figures holding hands represent community and togetherness.

Now try this

Find **two other** texts from pages 74 to 83 that use images and graphics.

Putting it into practice

In this section you have revised identifying and understanding:

- paragraphs and columns
- fonts
- titles and headings
- lists and bullet points
- tables and charts
- images and graphics.

Look at the extract from Text H and the test-style question below and read a student's answer.

Text H

Restaurant Work Experience Fact Sheet

You will need to wear:

Your chef's jacket

Sensible shoes or boots

You should bring:

- signed parental consent form
- completed health questionnaire
- £5 to cover food costs
- white shirt and black trousers
- notebook and pen.

For the third year running, Estrick College's catering students have been offered the chance to spend two weeks working at Estrick Hotel. This is a fantastic opportunity to get further knowledge about the catering industry and gain hands-on experience at a top-class hotel kitchen.

Worked example

Text H uses paragraphs to present information. List **two** other features of Text H that help to present information.

You do **not** need to write in sentences.

1 box

2 chef

(2 marks)

✗ The student has correctly spotted that the writing in the box and the chef are important features. However, the student's answer is not precise enough. A better answer would be **bullet points**.

✗ The student has identified what is in the picture but has not made it clear that it is an **image** of a chef. A good answer would be: **image of a chef**.

Now try this

Look at the whole of Text H on page 81. Identify **two other** presentation features you could use to answer the questions above.

Understanding detail

You should read both texts in your test carefully to make sure you understand details.

This includes finding individual details and selecting true statements from the text.

The questions on this page are about Text I on page 82 – you don't have to answer them, they are just examples of the types of question you might find in your test.

Finding information

Focus your reading on the key words to help you find the information to answer the question.

Look for the words **college courses** in the text.

1 How can you find out more about the college courses?

Look for the phrase **employers choose students** in the text.

The question asks for **one** answer.

2 Give **one** reason why employers choose students from the college.

Choosing the correct option

At least one of the detail questions will be multiple choice.

Make sure your answer matches the question.

Look for **numbers** in the text.

Look for the word **years** in the text.

3 The Building Trades Department of the college was set up:

A 5 years ago ☐

B 20 years ago ☐

C 3 years ago ☐

D 7 years ago ☐

Selecting true statements

Questions asking you to select true statements from the text are usually multiple choice. Read the options very carefully to make sure you choose the correct one. Here, B is false because the college provides all materials and tools.

If you are struggling to find the true statement in the list of options, start crossing out the ones you know are false. You will be left with the correct answer.

4 Which **one** of the statements about the college is true?

A Students need to have a grade C GCSE in Maths and English. ☐

B Students need to buy all materials and tools ☐

C Students need support with English, Maths and IT skills. ☐

D All students get a construction job within a month of leaving college. ☐

Go to pages 3 and 4 to revise how to answer the different types of question.

Now try this

For **each of the four** questions above, underline the key words that would help you to select the correct detail in the text.

Reading for detail

When you have finished skim reading the texts in your test, you should read them again carefully, looking for important details that you may be asked about.

Looking for details

Before you look for details in the text:

 Read the question carefully to find out what you need to look for.

 Skim read the text to find out where the information is likely to be: if the text has sub-headings, bullets or a table, look at those features first.

 Identify key words in the question and search the part of the text you think the information is likely to appear in.

Worked example

Give **one** reason why employers choose students from the college:

You do **not** need to write in sentences.

students are fully prepared for the workplace

(1 mark)

Text I

A quick skim read of Text I reveals two sub-headings that contain key words used in the question.

The first sub-heading gives reasons why the reader should choose Estrick College.

> We have invested in new facilities. Last year, a new building trades centre was added. It has workshops for bricklaying, joinery, painting and decorating, plastering, plumbing and electrical installation work. We aim to prepare students for life on a real construction site. We also built new classrooms and an assembly hall. This year, we have spent over £2 million on a state-of-the-art computer suite. These improvements will support students with the academic side of their college courses.
>
> **Why choose Estrick College?**
>
> - All our courses are industry-approved.
> - Our workshop facilities meet all health and safety standards.
> - We provide all materials and tools.
> - Our staff are fully qualified and friendly.
> - We offer support with English, Maths and IT skills.
> - Over 75% of students find a construction industry job within one month of leaving college.
>
> **Why do employers choose Estrick College students?**
>
> Our students are fully prepared for the workplace. All students have experience of working on real construction sites. For the past 3 years all our students have left the college with industry-approved qualifications in their particular trade subject.

The second sub-heading gives reasons why employers choose Estrick college. This means this section will be a good place to look for details.

Now try this

Read the whole of Text I on page 82 and answer the test-style question on the right.

Which **two** of the following has the college recently built?

A   The computer suite ☐
B   The gym ☐
C   The website ☐
D   The materials and tools ☐
E   Classrooms ☐

Careful reading

In your test, you should read both texts carefully to make sure you find exactly the right piece of information for each question.

Reading carefully

Statements in multiple choice questions may not be written exactly as they are in the text. This means you must read sections of the text very carefully to find the correct answer. Read the extract from Text A below and look at a student's answer to a test-style question.

Text A

Dear Editor,

Positive role models in the media

I have decided to write to you to complain about the lack of positive role models in your magazine.

Your magazine is for people between the ages of 13 and 18, who often feel under pressure to 'follow the crowd' so they can fit in with their friends. They look to your magazine for guidance. Your glamorous stories and photographs seem to be all about appearance, even though there is much more to life! You should be inspiring young people to be happy, healthy and ambitious.

Worked example

According to Text A, the magazine is bad for young people because:

A   it is for young teenage girls   ✱

B   the Academy is very disappointed   ☐

C   the magazine contains glamorous stories   ☐

D   there is a lack of positive role models   ✗

If you are struggling to find the answer in the text, you can start ruling out each of the options instead. You can also do this when you have answered a question to check that you are correct.

After a quick skim read, this answer might seem correct. The magazine is for young girls, but this is not the reason why it is bad for young people.

At first glance, you might think these answers are correct, as they both include exact words from the text. Reading the text carefully will show you that they are not.

You can find the answer in this part of the text. The first paragraph clearly states that the writer is complaining because there is a lack of positive role models for young people – this shows that D is the correct answer.

Now try this

Why do you need to read the question and text very carefully when answering multiple choice questions?

Tricky questions

Read the whole question carefully before answering to avoid any misunderstandings.

Tricky multiple choice questions

Some of the wrong answers to the multiple choice questions may look like right answers. You should:

- read the question again carefully
- think carefully about each of the answers
- cross out those that are definitely wrong.

This should leave you with the correct answer. You can also do this at the end of the test to check that you have put the correct answer.

Tricky short response questions

When answering tricky short response questions you should:

- use only information you can find in the text
- focus on the information you need to answer the question – avoid getting distracted and reading too much into the text
- go back at the end of the test to check your answer.

Worked example

1 According to Text H, which **one** of these statements about the Estrick Hotel is true?

A It was built in 1920 ☐

B It has served afternoon teas in the Conservatory for 100 years ☐

C It has a snack bar in the grounds ☒

D The bar staff are students ☐

It is easy to misread dates, so make sure you read all sentences containing numbers very carefully.

The Conservatory Restaurant does serve evening meals, but it is the café that serves afternoon teas.

This answer is the only one left. Reading the text again shows that it is the correct one.

If your eye quickly finds the words **Bar staff** and **students**, you might choose this answer. But reading carefully reveals only students over the age of 18 can work in the bar. It doesn't say that all bar staff are students.

Worked example

2 According to Text H, give **one** reason why working at the hotel would be a fantastic opportunity.

to get further knowledge about the catering industry and gain hands-on experience

The words **fantastic opportunity** appear in the first paragraph of the text. There are **two** correct answers here but the student needs to give only **one** to get the mark.

Go to page 81 to read Text H.

Now try this

Make a list of things you should do in your test if you come across a tricky question.

Vocabulary

If you come across an unfamiliar word in your test, don't panic! This page describes some techniques that you can use in your test to help you work out the meaning of a word. You can also use a dictionary in your test, but remember this takes time.

Unfamiliar words

There are several things you can do to work out the meaning of an unfamiliar word:

Practise using these techniques when you come across a word you don't know in this book, or any other texts you read.

 1 Read the text before and after the word, looking for clues.

2 Think about whether it looks like any other words you know the meaning of.

3 Look at any images near the text.

Go to page 31 for useful tips on using a dictionary.

4 Look it up in a dictionary.

The picture shows devices that run on electricity.

The word **appliances** may be new to you. You can use the methods above to work out what it means.

It's something to do with electrics.

The phrase **Turn off appliances when not in use** tells you that appliances are something that you can turn on and off.

Electrical appliances cause 55% of all fires in the home.

• Keep all electrical appliances away from children.
• Turn off appliances when not in use.
• Do not use electrical appliances in bathrooms or near kitchen sinks – electricity and water do not mix!

The picture shows a fire.

You might not be familiar with the word **flammable**.

It is used in a sentence about matches and other firelighting equipment, so it is likely to mean something that can catch fire easily.

It looks like the word **flame**.

Open fires can spark and set furniture and carpets alight.

• Keep all matches and other fire-lighting equipment and flammable materials away from children.
• Do not leave a fire unattended.
• Keep all chemicals and aerosol products away from flames.

Now try this

Read the section about dangerous chemicals in Text C on page 76. Use the techniques above to work out the meaning of the word **ingested**.

Using a dictionary

You may already know how to use a dictionary, but here is a reminder.

Using a dictionary

Dictionaries are easy to use if you follow these rules.

- The words are in alphabetical order, so all the words beginning with **a** will be together.
- After finding the section for the first letter of the word, you need to look at the second letter – so **apple** will come before **arrow** as **p** is before **r** in the alphabet.
- The words are in bold to make them easy to find on the page.

Getting it right

Don't spend too long looking up words in the dictionary. Always try to work out the meaning of unfamiliar words using the method on page 30 first. You could practise using a dictionary before your test by looking up any unfamiliar words you come across in this book.

Understanding a dictionary entry

This is the word you are looking up.

This is the definition, or meaning of the word – sometimes there is more than one meaning.

Dictionaries often show words that have a similar meaning to the word you have looked up. Looking at these can be the quickest way to work out a word's meaning.

> **dictionary** *noun*
>
> a book or electronic resource that lists words in alphabetical order and gives their meaning and examples of similar words
>
> *Similar words:* vocabulary list, vocabulary, word list, wordfinder

Where to start looking

Look at the tricky vocabulary used in the warning notice (from Text D) on the right. Follow the steps below to find the definition of **external** in the dictionary.

1. The word **external** begins with **e**, so you should look in the **e** section of the dictionary.

2. The second letter is **x**, which comes at the end of the dictionary, so the word will be found towards the end of the **e** section.

> **WARNING TO PARENTS OR GUARDIANS**
>
> The product is safe for external use only. If swallowed, product could be a throat irritant: you should seek medical help immediately. If heated, vapours can cause headaches.

> **external** *adjective*
>
> part of or forming the outer surface of something: 'the external walls'
>
> *Similar words:* outer, outside, outermost, outward, surface

Ex- is the prefix of many words that mean **out** or **from**, such as 'exit' and 'expel'.

Some dictionaries give you an example of the word in a sentence.

Now try this

Use a dictionary to find the meaning of the word **vapours** from the text above.

Putting it into practice

In this section you have revised:

- reading for detail
- reading carefully
- answering tricky questions
- dealing with unfamiliar vocabulary
- using a dictionary.

Look at the test-style questions below and read a student's answers.

Getting it right

Remember, before you start:

- read the questions and pick out the key words
- skim read the texts and find the main ideas
- underline useful parts of the text.

Worked example

1 According to Text D, what is it safe to do with the worms?

A leave them to dry ☒

B register your worms ☐

C swallow your worms ☐

D hold them in your hand ☐

(1 mark)

✗ The student has not read the right section of the text. The key word in the question is **safe**.

Go to page 77 to read the whole of Text D.

Worked example

2 According to Text D, list **one** thing you can do on the worm website.

enter the size and post photos of your worms

(1 mark)

✗ The question asks for **one** point and is worth 1 mark. The student has given one correct answer and one incorrect answer.

Worked example

3 According to Text D, how do you contact the company if you have a problem with the worm farm?

on the website

(1 mark)

Contact information is usually easy to spot. Look out for a phone number, email address, or address.

✗ The text doesn't say that you should contact the website if you have a problem with the worm farm. The student has not used the correct information in the text to answer the question.

Now try this

Read Text D on page 77 and then answer the above questions correctly.

Using information

For your reading test, you will need to show that you understand how to use the information found in a text. This means reading carefully and selecting information for a particular use, or purpose.

How to use information

We can use information in texts for a variety of reasons.

- Descriptive, persuasive and informative words and phrases can be used to persuade people to do something.

- Informative and instructive texts can be used to help or advise the reader.

Read about the purpose of using information on pages 15 to 18.

Worked example

1 Safety in the home is important.

Using Text C, find **two** things that we should do in the home to keep children safe.

You do **not** need to write in sentences.

1 Keep cleaning products away from them
2 Keep matches away from them

(2 marks)

The information in this text can be used to inform people how to keep safe at home.

Text C

- Keep all matches and other fire-lighting equipment and flammable materials away from children.
- Do not leave a fire unattended.
- Use a fireguard at all times.
- Keep all chemicals and aerosol products away from flames.

Many household products contain dangerous chemicals. They can cause serious illness, or even death, if ingested. They can also cause burns if they come into contact with the skin.

- Keep all cleaning products away from children.

Worked example

2 You want some friends to join you at the nature reserve.

Using the information in Text F, give **two** things that would convince your friends that Estrick Nature Reserve's accommodation is comfortable and well-equipped.

You do **not** need to write in sentences.

1 hot tub on terrace
2 fully-equipped kitchen

(2 marks)

Text F

I was lucky enough to stay in the largest and most luxurious property offered by Estrick Nature Reserve. **Longacre Lodge** is the original hunting lodge which has been lovingly restored and now offers superb accommodation for up to twelve people. It even has a hot tub on the terrace! Like all the holiday properties at the reserve, it comes with a fully-equipped kitchen and a welcome pack is provided on arrival.

The information in this text can be used to persuade someone to go to the Estrick Nature Reserve.

Now try this

Read the whole of the nature reserve article on page 79. Find **one more** reason why someone would find the accommodation comfortable and well-equipped.

Responding to a text

For your test you will also need to show that you understand how to respond to a text. You will need to read the texts carefully and think about how the writer wants you to respond.

Getting it right

To answer questions about responding to a text, you should follow the same steps as for other questions.

- Read the questions carefully and underline key words.
- Skim read the text to find the best places to look for information.
- Go back and read the text carefully to make sure you find the correct answer.

Worked example

1 Your friend wants to stay at the nature reserve but is worried that <u>visitors will harm the wildlife on the reserve</u>.

Using Text F, give **two** reasons to convince your friend that visitors are <u>helping wildlife on the reserve</u>.

1 visitors can help to plant trees
2 visitors are involved by checking on feeding boxes

(2 marks)

The writer of Text F describes how visitors can help at Estrick Nature Reserve to encourage them to visit.

> Go to page 79 to read Text F.

Worked example

2 Your friend is not convinced that it is a good idea to keep the online donation feature on the animal shelter website.

Using Text J, give **two** facts to reassure your friend that the online donation feature is a good idea.

1 people don't have to leave home to donate
2 online donations are higher than those put into collecting boxes

(2 marks)

The writer of Text J informs the audience about the online donations feature. They describe the benefits of the online donation feature in order to encourage the audience to use it.

> Go to page 83 to read Text J.

Now try this

Read the whole of Text J on page 83. List **three** ways a reader could respond to the text by getting involved in fundraising.

Putting it into practice

The last two pages have helped you to revise:

- using the information in a text
- responding to the information in a text.

Read the test-style questions below and look at how a student has answered them.

Text G

Taking up the clippers!

Posted on 27th April

A customer asked me today how she could become a hairdresser. I told her it is very hard work! You are on your feet all day. The constant hair-washing can make your hands dry. You may not like all of the customers as some may be rude, and some may even refuse to pay. Also, you will always have to work at weekends. But last week I styled the hair of three local girls before their school prom. I had been cutting their hair since they were three years old. Seeing them look so beautiful made the sore feet so worthwhile!

✓ The student has looked for the key words **hard work** in the text and has come across one answer. There are several answers in the following section as well.

✗ The student has found the correct part of the text for this answer but has not read it carefully. The text does not say that customers hate hairdressers!

Worked example

1 Your friend does not think being a hairdresser is hard work.

Find **two** negative things from Text G that would show your friend that hairdressing **is** hard work.

1 you are on your feet all day
2 customers hate hairdressers

✓ This answer is correct. It is one way in which the salon is suitable for children.

✗ The student has not given enough information in their answer.
Careful reading of the text shows that **top-quality** hairdryers and scissors are used. It is this information that could persuade your friend to go to Kids' Cutters.

Worked example

2 Find two pieces of information in Text G to convince your friend that Kids' Cutters is a high-quality hairdressers that is suitable for children.

1 additive-free shampoo
2 hairdryers and scissors

Now try this

Read the whole of Text G on page 80 and find the correct answers to the questions above.

Avoiding common mistakes

During your test you will need to stay calm to avoid making simple mistakes. Read the tips below and look at a student's answer to test-style questions to help you to avoid common mistakes.

Using the right information

It is important that you use only the texts provided in the test. The text might be on a topic you know a lot about, but you should not use your own knowledge. If the information is not in the text, don't use it!

Writing the right amount

You don't need to write your answer in sentences. Don't waste time writing full sentences if a question does not ask for them.

Reading the question carefully

It is important not to rush your answers. Always read the questions carefully first, to make sure you find the correct part of the text.

> Careful checking of your answers at the end of the test will help you to catch mistakes if you have made any.

✓ The first answer can be found in Text D and is correct.

✗ The second answer does not come from Text D. It might be true, but if the information is not actually in the text, you won't get the mark.

✗ The student has written a full sentence, which you don't need to do.

Worked example

1 Your friend thinks that his son will find the worm farm boring.

Using the information in Text D, give **two** points to convince your friend that his son will not find the worm farm boring.

You do **not** have to write in sentences.

1 you can share the worms with friends
2 His son won't find it boring as the worms are great for boys who like nature.

(2 marks)

✗ The student has rushed into an answer without reading the question or the text carefully. The question asks about faults with the product – it doesn't ask what you should do if the product is used incorrectly.

Worked example

2 According to Text D, what should you do if the product is faulty?

You do **not** have to write in sentences.

Seek medical help immediately

(1 mark)

Now try this

Read the questions on this page carefully. Underline key words in the questions. Then read Text D on page 77 carefully and correct the mistakes the student has made.

Checking your work

When you have finished answering the questions in your reading test, it is a good idea to go back over your answers to check for mistakes. This page will help you to spot some common mistakes.

 Writing a good answer

Be careful that your answers are not too short, or too general.

✗ The first answer is too general. Students will make food, but the text is more specific than this.

✗ The second answer is too short. The student has not focused on the word **activities** in the question.

Worked example

1 Identify **two** activities from Text H that students will be able to do during their work experience.

You do **not** need to write in sentences.

1 making food
2 sandwiches and drinks

 Reading the question carefully

If a question asks about specific details, such as contact details, make sure you read the question and text carefully.

✗ This student has written the website address you need to use to register interest in the work experience. It is incorrect because the question asks **how** you can register.

Worked example

2 According to Text H, how do you register your interest in work experience at the hotel?

You do **not** need to write in sentences.

www.estrickhighschool.com

 Tricky multiple choice questions

Make sure you give the correct number of answers. For multiple choice questions, you could be asked to select one or two options. If you select two options when you have been asked for one, you will not get the mark.

Getting it right

Always underline the key words in the question. It will help when you check your answers at the end of the test.

It is a good idea to keep five minutes free to check your answers at the end of the test.

Now try this

Read the questions on this page carefully. Underline key words in the questions. Then read Text H on page 81 and correct the mistakes the student has made.

Writing test skills

For your Functional Skills writing test you will be given two writing tasks. These will test your writing skills, including spelling and punctuation.

Good writing skills

To do well in your writing test you need to:

✓ write clearly

✓ use details

✓ present information in a logical order

✓ use correct grammar, spelling and punctuation

✓ use suitable language for your audience

✓ use a suitable type of text and layout for the purpose.

Writing skills in use

Look at Texts D and J to see examples of good writing skills.

If you are writing instructions, a numbered list will help readers to follow them in the right order.

The writer has developed the instructions with extra detail about what could go wrong. This is important information, as the product will be used by children.

Here, the writer has used an apostrophe and an exclamation mark correctly. It is important to use the correct punctuation to express information so that the audience understand the meaning.

Here, the language is formal because the writer does not know the audience.

The writer has used new paragraphs for each fundraising event.

Go to pages 52–56 to revise structuring your writing.

Text D

1. Fill a large plastic bowl with water. Be careful not to use hot water as this might make your worms grow too quickly.
2. Put the worm mixture into the water. Don't worry if it sinks to the bottom in one big lump – it will soon start to look like worms!
3. Follow the pictures on the enclosed instruction leaflet to put the farm together.
4. After 24 hours (one whole day and night) remove the worms from the water. Put them on pieces of kitchen paper and leave them to dry for an hour.

Text J

For the second year running, our Fun Run was held during a torrential downpour. This did not dampen the spirits of the runners, who managed to raise a total of £1,000 in sponsorship! This event is growing in popularity and we will need more volunteers along the route next year.

Unfortunately, the car wash event was poorly attended. The Scouts did an excellent job but we will need to ask for more volunteers next year to advertise this event. The same goes for the Bring and Buy sale, where we need people to ask local companies to donate some exciting products.

Now try this

1 Why do you need to include details in your writing?
2 List **two** ways writers can present their information in a logical order.

Writing test tasks

For your writing test, you will need to complete two writing tasks. You should read each task carefully to make sure you know exactly what you are being asked to do.

Before you begin writing your answer:
- read the task carefully
- read the information section
- identify the audience, purpose and format
- plan your answer.

Writing task information

Look at the examples of writing tasks below, based on Task A (page 84) and Task G (page 90).

You should read the writing task and all the information provided carefully. The task may give you bullet points.

These can be used to help with ideas and structure. You can use the details in the information section for the first three bullet points.

Task 1

Information

You bought a set of pens from Estrick Stationery Supplies. The first time you put the pens into your bag, one of them leaked onto your sports kit.

When you returned to the shop the assistant was rude and refused to refund your money for the pens.

Writing task

Write a letter of complaint to Martin Jones, Manager of Estrick Stationery Supplies Ltd, 8 High Street, Estrick, EW12 5HH.

You may wish to include:
- the product details, including where and when you bought the pens
- what went wrong with the pens
- how you have been treated by the shop assistant
- what you want to happen now.

Use sentences and write in Standard English.

(15 marks)

Task 2

Information

You read a selection of posts on an internet discussion about recycling.

Writing task

Write your own message in response to this internet discussion, giving your detailed views on the subject of recycling.

Use sentences and write in Standard English.

(10 marks)

Here, the task is to write a message on an internet forum, giving your views about the topic. This tells you that:
- **you can use some informal language**
- **you will need to develop your points with detail.**

You should read the other posts carefully so that you can respond to each of the points already made.

Now try this

Look at the sample Task A above. The audience, format and purpose have been underlined. What do they tell you about how to approach the task? Make some planning notes like the bullet points above.

Putting it into practice

You now know what to expect from your writing tasks and what skills they will test.

Read the tasks below and look at how a student has prepared. These test-style tasks are extracts from Task C (page 86) and Task H (page 91).

Task 1

Information

You noticed this letter in your local newspaper.

Estrick News 20th January 2016

Letter of the week

Dear Editor,

I have noticed that most young people seem to spend all day on their mobile phones. Why? What is so interesting that they need to ignore what is happening around them?

Last week a young man bumped into me on the pavement because he didn't look up from his phone. I could have fallen and hurt myself badly.

I think young people should only be allowed to use their phones for two hours a day. They should also not be allowed to use their phones public.

Yours sincerely,

Anonymous

Writing task

Write an <u>article for the newspaper</u>, giving your views on the subject of this letter.

In your article, you should:
- say whether you agree or disagree with the letter
- say why you agree or disagree
- give reasons that support your views.

You can add any other interesting information.

(15 marks)

Task 2

Information

You and your friends recently had a day out at Estrick Nature Reserve.

ESTRICK NATURE RESERVE

Have fun while you learn to enjoy nature!

- Small animal zoo
- Children's adventure playground
- Nature trail with experienced guides
- Bird-watching area
- Boating lake
- Café, restaurant and gift shop
- Climbing wall and high-ropes activity centre

Writing task

Write an <u>email to your friend</u>, telling them about your day out at Estrick Nature Reserve.

In your email you may:
- give details of your visit
- explain what you liked or disliked about it
- say why it would be a good or a bad place to visit.

(10 marks)

Go to page 1 for more on planning your time in the test.

Sample answer

Task 1 – <u>30 minutes</u>, use titles, sub-headings and formal language, use detailed evidence to disagree with letter

Task 2 – <u>10 minutes</u>, no 'Dear' needed for email

✗ The student has not included any time for planning or checking and has not divided the time between the tasks correctly. This will mean Task 2 may be rushed or unfinished.

✓ The student has made some useful planning notes about audience, format and purpose.

✗ The student has not underlined anything in the task to help with purpose.

Now try this

Correct the mistakes the student has made with timings for the above tasks. Then read Task H carefully and make planning notes about audience, format and purpose.

Understanding audience

You will need to make sure that your writing is suitable for its audience. **The audience** is the person who reads a text. It is important to think about what your audience needs to know.

Who is the audience?

In the test, the tasks will make the audience clear. You should think about your audience in terms of age, interests, occupation, and likes and dislikes.

> Knowing who the audience is will help you to choose appropriate vocabulary and presentation features.

Task 1

INFORMATION

A group of international exchange students is coming to your town to spend a month learning English. They are keen to visit all the attractions in your local area. You decide to let them know more about your local area by writing about what there is to do and see. You have heard they like sport, eating out and walking.

WRITING TASK

Write an article for your school/college/workplace newspaper, describing the kinds of things the international exchange students can do in your area.

In your article you may want to include details of:

* sporting attractions
* restaurants
* places to walk, such as local countryside.

> This test-style Task 1 is an extract from Task D on page 87. For this task, you have a choice of audience. What information you choose will depend on your audience.
>
> The audience here is the reader of the newspaper.

> You might find it useful to choose an audience that you can relate to. This will make it easier to think about what the audience like doing.

Task 2

INFORMATION

You are a member of a group about to take part in a sponsored swim to raise money for a local charity. You receive an email from the group's leader asking you to arrange refreshments. Write an email to Maya Patel owner of Estrick Eats.

In your email you should:

* give details of the sponsored swim
* persuade her to provide refreshments.

> This is a test-style Task 2 based on Task A on page 84.

> The bullet points in the writing task give ideas for your writing. You should follow these, but think carefully about what information your audience needs.

Sample answer

Audience – Maya Patel, owner of café

To persuade her to provide refreshments, Email should by formal and contain details of sponsored swim

✓ The student has identified the audience, purpose and type of text in their plan.

✓ The student has realised that the email should be formal.

Now try this

Read the full version of Task D on page 87. Choose one of the audiences and make some planning notes about what information they would need.

Letters and articles

In your writing test, you could be asked to write a letter or an article.

Writing a letter

You need to know when and how to write an informal and formal letter. You would write a formal letter to someone you don't know, or to an official person or organisation. You would write an informal letter to friends or family, or someone you know well.

A letter should include:

- your address at the top right of the letter
- the address of the person you are writing to on the left and lower down
- the date
- Dear Sir/Madam (if you don't know the person's name)
- Dear Mr/Ms (if you know the person you are sending the letter to)
- Hi... (if you are being informal)
- a heading to draw the reader's attention to the topic
- Yours sincerely followed by a comma (if you are being formal)
- Thanks... (if you are being informal)
- a new paragraph for each new point you want to write about.

18 Wool Street
Nottingham
NG2 2JH

4th February 2016

A Sharif,
Manager
Estrick Stationery Supplies Ltd
8 High Street
Estrick
EW12 5HH

Dear Ms Sharif

Faulty pens

Yours sincerely,

Mary Smith

Writing an article

The main purpose of articles is usually to describe, but they can also be used to inform and to persuade.

An **article** should include:

- short headings and sub-headings to grab the reader's attention and tell them what the article is about
- paragraphs to break up the text into easy-to-read sections
- facts and opinions to inform the reader.

It is important to remember capital letters at the start of headings and sub-headings.

Mobile phone danger?

Phones for safety

Phones for information

Go to page 22 for more about using headings and sub-headings.

Now try this

1 Name **three** features that you must include when writing a letter.
2 What should you remember when writing a heading for an article?

Emails and online discussions

In your writing test, you may be asked to write an email or an internet forum post. This page describes the main features you should include in these types of text.

Emails

Like letters, emails can be formal or informal. They tend to be informal if you know the person you are writing to, and formal if you are writing to someone you don't know.

For an **email**, make sure you add a heading line, and 'Dear' at the start. If you are writing to a friend or family member, you can start your email with 'Hi'.

Emails can end in a more informal way than letters. You can end with 'Thanks' if you are asking for something or 'Regards' if you don't know the person you are writing to.

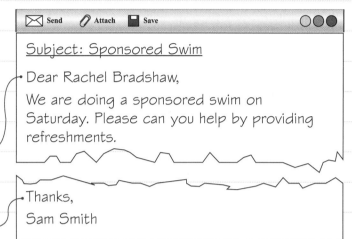

⊠ Send   ⁄ Attach   ▣ Save   ○○●

Subject: Sponsored Swim

Dear Rachel Bradshaw,

We are doing a sponsored swim on Saturday. Please can you help by providing refreshments.

Thanks,
Sam Smith

Internet forum discussions

Internet forum discussions are usually informal.

When writing a response to an **internet forum discussion** you should include:

* your name on the left
* the date on the right
* your opinion on the topic of the discussion
* responses to the main points in previous messages
* any questions you have that relate to the topic.

forum *noun*

a website where people can share their ideas and discuss topics or issues

Getting it right

Before your test, practise the correct way to write a date on a letter, email or online discussion. The day comes first, then the month, then the year. Remember to use a capital letter for the month.

Look at Task G on page 90 to see the typical features of an internet forum.

Now try this

1 When can you be more informal when writing an email?
2 Make yourself a revision table like the one below so that you know the correct layout for each format.

Format	Layout
Letters	
Articles	
Emails	
Internet forum discussions	

Formal writing

You should read the writing task information carefully to work out how formal your writing should be.

Letters and articles will usually need to use a formal writing style. Emails and posts on online forums can be informal.

When to use formal writing

You should use a formal style if:

- you are applying for a job
- you do not know your audience well
- your purpose is official, for example making a complaint
- you are writing to somebody official, for example a recruitment agent
- you are writing and article or a letter to someone you don't know.

Using Standard English

When writing a formal text, you should use Standard English, which means using:

- complete words (I am)
- complete sentences
- correction punctuation and grammar.

You should avoid:

- text language, e.g. 'LOL'
- slang, e.g. 'I was gutted'
- contractions, e.g. 'I'm', 'don't'.

Writing a formal letter

Read the test-style task extract on the right, based on Task B (page 85) and look at a student's answer below.

> Go to pages 59–63 to revise sentences and grammar.

Task 1

Write a letter to Jane Edwards, Head of the Environmental Department, Estrick Council, Main Street, Estrick, ES1 3MS to inform her of the problem with the community centre roof.

In your letter you should:

- introduce yourself
- describe the problem
- explain who could be at risk
- say what you want the council to do about it.

Remember to set out your letter correctly using the correct layout.

If you do not know if a woman is **Miss** or **Mrs** you should use **Ms**. Never use a first name in a formal letter.

Notice how 'I am' is written in full. —

Complete sentences are used. —

If you are handwriting a letter, you should start a new paragraph on a new line and write the first word further to the right.

If you are typing a letter, leave a blank space between the paragraphs.

Sample answer

Dear Ms Edwards,

I am writing to inform you about a safety issue in the local park.

I have noticed that several slates on the roof of the community centre are loose. Last week some of the slates fell on to the path in front of the entrance. This is a serious safety problem because many people use this path.

Now try this

Read the information for the task above on page 85. Write **two more** sentences of the letter, using formal language.

Informal writing

In your writing test, you need to work out whether your answer should be formal or informal. This page explains when and how you should use informal writing.

When to use informal writing

You can use an informal writing style if:

- you know your audience personally
- your purpose is to be friendly, for instance if you are asking for help or sponsorship
- you are adding your views to an informal discussion
- you are using an informal format, like an email or an internet forum discussion.

Writing for an audience you know personally

When using informal writing, you should use:

- informal contractions, e.g. 'don't' or 'can't'
- informal phrases
- correct grammar and punctuation
- correct spelling
- complete sentences
- specific text features.

Writing informal emails

When the email audience is a friend, the greeting can be informal, for example 'Hi'. For people you do not know, it is better to use 'Dear'.

Look at the test-style task on the right, based on Task H on page 91.

Notice that this email is informal, but it still uses complete sentences.

The student has used informal language here. If you are using slang words or phrases, make sure you know the reader will understand them.

Always remember to sign off a letter or email. When the email audiences a friend, you can sign off with 'Thanks,' or 'Regards,'.

Exclamation marks are a good way to show excitement in an informal format like an email. Don't use too many as they will lose their impact.

Task 2

Write an email to your friend, telling them about your day out at Estrick Nature Reserve.

In your email you may:

- explain what you liked **or** disliked about Estrick Nature Reserve
- say why you think it would be a good **or** a bad place to visit.

Sample answer

Hi Dan, you'll love the Nature Reserve. The high ropes are a total gas, but most importantly for you, the café food is out of this world!

Regards,

Beth

Getting it right

You should still use correct spellings, grammar and punctuation in informal texts.

Now try this

Read the information for the task above on page 91. Then write **two more** sentences of the email, using an informal writing style.

Putting it into practice

In your writing test, you will need to show that you understand:

- the audience
- the type of text
- the purpose
- formal and informal texts.

Read the test-style task below, based on Task D on page 87, and look at how a student has answered.

Getting it right

If a question gives you a choice of audience, make it clear which audience you have chosen. You can do this by using information that suits your chosen audience.

The information in an article should sound reliable and trustworthy. Formal writing is the best way to do this.

✓ The correct format has been used for an article. The student has included headings, sub-headings and paragraphs.

✗ The audience of the article is not clear. The student seems to be writing for both students and parents.

✗ The student has used informal phrases. They would be acceptable in an email to a friend, but are not suitable for an article.

✗ The student has used the wrong style. Words like kids and rave are too informal for an article, as you do not know the audience personally. The student has also used slang like till and contractions like you'll, which are not formal enough for a serious article.

Task 1

Write an <u>article</u> for your school/college/<u>workplace newspaper</u>, describing the kinds of things foreign exchange students can do in your area.

In your article you may want to write about:

- <u>local attractions</u>
- <u>leisure facilities</u>
- shopping and local markets.

You could add any other useful information.

Sample answer

<u>Local attractions</u>

Estrick has loads of historic stuff to visit. The castle has walls that are five foot thick. The kids will get totally <u>wiped out</u> walking all round them! They do history talks at the weekends which can be a real help with your homework. It's really cheap to get in and students get discount.

<u>Leisure facilities</u>

Estrick has a superb new leisure complex right in the middle of town. You'll love the Olympic-size swimming pool and the <u>kids</u> will <u>rave</u> about the slides. Just wait till the waves start! There's a café where you can get wicked burgers and chips after all that exercise in the pool. It really does make our little town worth visiting!

Now try this

Read the information for the task above on page 87. Rewrite the student's answer.

1 Use the correct formal style.
2 Make the audience clear.

 If you decide to write for a workplace newspaper, you will need to change the information about homework.

Understanding planning

Before you start writing, it is a good idea to make a plan. This will help you to:
* make sure your writing suits the audience, purpose and format
* develop your ideas with the right amount of detail.

Informed planning

Before you start planning, read the task and the information section carefully. Some tasks have bullet points that you can use as a starting point for your plan.

In your plan you should include brief notes on the audience and purpose, and any key information you want to include. Don't spend too much time writing your plan. You can put all the detail you want in your answer.

Writing a good plan

There are different ways you can plan your writing. Choose one that suits you. Look at how a student has used information in the task below to help them plan their answer to Task H on page 91.

You can tick off each idea in your plan as you write to check that you have covered everything.

On the online test you can use the notepad to write planning notes. Any notes you make here will not be marked.

Task 2

Information

You and your friends recently had a day out at Estrick Nature Reserve.

ESTRICK NATURE RESERVE

Have fun while you learn to enjoy nature!

* Small animal zoo
* Children's adventure playground
* Nature trail with experienced guides
* Bird-watching area
* Boating lake
* Café, restaurant and gift shop
* Climbing wall and high-ropes activity centre

Writing task

Write an email to your friend, telling them about your day out at Estrick Nature Reserve.

In your email you may:
* explain what you liked **or** disliked about it.
* say why it would be a good **or** a bad place to visit.

Getting it right

In the test, you can plan what you are going to write on the test paper. Remember:
* When you are finished, cross your plan out neatly to show that you don't want it marked.
* Don't waste time writing a detailed plan. Write key words that will remind you of your main ideas.

Sample answer

Estrick Nature Reserve
* liked the reserve, lots to do
* good places to eat ✓
* playground really good for children
* high ropes are fun for teenagers

Now try this

Look at Task B on page 85. Using one of the methods above, use the bullet points in the task to start a plan that will help you to answer the question.

Using detail

All the writing tasks start with an information section. You should use this information to add detail to your writing. This will help you to give your audience all the information they need.

Where to use detail

Read the information section to help you think of ideas for your writing. If the task has bullet points, read them and then skim the information section for relevant details.

You should also use your imagination to think of details you can add to your writing to make it interesting to read.

If you are asked to write about a topic you don't know much about, don't panic! Use the information section to help you.

Go to page 43 to revise writing for internet forums and blogs.

Adding detail to your points

Read Task G on page 90 and look at a student's plan and detailed answer on the right.

Notice how this plan uses the information in the task. The student has decided to agree with Ben, so the plan answers the points made by Jake.

Task G asks for your detailed views about recycling. The student has included the main points from their plan and added additional detail to help make their writing more descriptive and persuasive.

Sample answer

Plan
- Audience: Adult
- Forum discussion: informal
- Purpose: To describe
- Agree with Ben
- Ask council for free bin
- Everything does get recycled - good for environment

I agree with you, Ben. Unless we recycle we will lose our countryside to landfill. You should ask the council for a bin, Jake. They provide them free of charge. There is no need to take anything to the tip when you have a bin. The council will take all your recycling and the council do recycle it all. New stuff is made from the recycling. This raises money for the council which covers the cost of transport, fuel and staff. This keeps your council tax down!

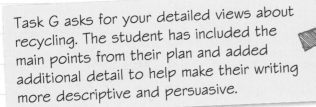

Now try this

Look at Task I on page 92. Use the information in the task to plan your answer.

Ideas

The writing task may use pictures to provide you with information. You should use these to help you plan your answer. You will need to use your imagination to add ideas and detail to your plan. If the task has pictures you can use them for ideas. Read the test-style task below, which is an extract from Task B on page 85. Look at the picture and see what extra detail it provides.

Task 1

Information

Walking through your local park, you notice the poor condition of the roof of the community centre. A large number of slates are loose and you are worried that they will fall on to the path and hit somebody. This could be very dangerous because there are always a lot of young children in the park.

Writing task

Write a letter to Jane Edwards, Head of the Environment Department, Estrick Council, Main Street, Estrick, ES1 3MS, to inform her of this problem.

In your letter you should:
- introduce yourself ✓
- describe the problem ✓
- explain who could be at risk ✓
- say what you want the council to do about it. ✓

Remember to set out your letter using the correct layout.

Sample answer

Plan
- Audience: Head of the council (adult) – needs to be formal
- Purpose: to inform and persuade
- Letter: use addresses, date and name
- Key ideas:
  - Loose slates, heavy and sharp, dangerous
  - children could get hurt
  - urgent fix needed

Notice that the student has used the bullet points from the task to inform their plan. The student has added extra details using details from the image.

Go to page 42 to revise how to set out a letter correctly.

Getting it right

If there are images in the writing task, look at them carefully. Images can help you think of descriptive language that you could include in your writing.

Now try this

Look at the plan you made on page 47 for Task B (page 85). Add extra detail to your plan using the information on this page and any ideas you get from the photo.

49

Putting it in order

When you have finished your plan, you will need to put your main points and ideas into a logical order. The most important information should come first.

Deciding on an order

In your writing test, you should write your ideas and points in order of importance. To avoid having to cross out your work, you should decide on the order of the ideas in your planning.

Planning tips

Begin with the most important idea and start a new paragraph for each new idea or point.

Don't be afraid to cross things out and change the order of your plan. It is easier to change your plan than to change your writing!

Planning an order

Read this test-style task based on Task C on page 86, and look at a student's plan below. The student has numbered their main ideas in the order they want them to appear in their answer.

Task 1

INFORMATION

You noticed a letter in your local newspaper.

WRITING TASK

Write an article for the newspaper, giving your views on the subject of this letter.

In your article, you should:
- state whether you agree or disagree with the letter
- say why you agree or disagree
- give reasons that support your views.

You can add any other interesting information.

Go to page 87 to read the rest of Task C.

Sample answer

Plan

- Audience: Adult
- Purpose: to inform and persuade
- Formal
- Disagree with letter
1. Phones are important
2. Why phones are important
3. Communicate with friends
4. Safety
5. Play games
6. Conclusion - phones are an important part of life - but young people don't spend all day on them

The student has used the bullets in the writing task to help her organise her ideas into order.

Points 3-5 could form one parargraph and could appear in any order.

Now try this

Go back to the plan you did on pages 47 and 50 for Task B on page 85. Think about the order of your points. Number your details to make sure they are in order of importance.

Putting it into practice

To do well in your writing test you will need to plan your answer. Look at the test-style task below, based on Task 1 on page 92, and read the student's plan.

Task 2

Information

You noticed this letter in your local newspaper.

Writing task

Write an email to editor@estricknews.org, giving your views on the letter.

You may wish to include:

- where and when you saw the letter
- your views on the content of the letter.

Estrick News
1 August 2015
Letter of the week

Dear Editor,

I am writing to thank the young people who helped me when I fell over last week.

I was out shopping and was frightened by the large group of teenagers. They were listening to loud music and some of them had tattoos and piercings.

However, when I fell they immediately turned off the music and came to help me. None of the older people even stopped.

Kindness is very important to me. Thank you, whoever you are. I now know that there are some kind and helpful young people out there.

Yours sincerely,

Agnes Mencher

Estrick

Sample answer

Plan

- Audience: adult
- Formal email - needs name and subject line
- Tattoos and piercings aren't bad
- Not all teenagers are unhelpful
- Not surprised that teenager was kind

✓ The plan has notes on audience, purpose and format.

✗ There are no notes in the plan for the first task bullet point.

✗ The second task bullet point has been used, but not all the points in the letter have been used.

✗ There is no logical order to this plan. The most important point should be first.

Now try this

Finish the plan above by:

- planning your ideas for each of the task bullet points
- adding extra details to the points
- numbering the points on the plan.

When you have finished your plan, re-read the task information. Make sure:

- you have used all the information that has been provided
- your plan includes all the information your audience will need.

Using paragraphs

For both your writing tasks you will need to show that you can structure your work.

One way to structure your writing is to use paragraphs. A paragraph is a group of sentences about one topic or idea. By grouping your ideas into separate paragraphs, you can develop each point with detail. Letters and articles always need paragraphs.

Planning the structure of a text

Look at a student's plan (below) and answer (on the right) for Task A on page 84.

audience – adult, formal

use paragraphs and name/address

Paragraph 1 Reason for writing:
- Estrick Stationery Supplies
- last week
- set of 24 coloured gel pens

Paragraph 2 What happened:
- pens leaked – put inside bag but still in plastic packet
- inside of bag ruined and ink on sports kit
- no warning on pens about ink

Paragraph 3 What the shop did:
- assistant rude and unhelpful
- had receipt but no refund given
- said I must have left top off pen, but I had not used the pens yet

Paragraph 4 Action needed:
- immediate refund
- apology from assistant
- some money for new kit and bag

This is a detailed plan with a logical order. The plan has been used to write four paragraphs.

Each paragraph is about a different idea or topic.

Sample answer

I am writing to complain about a set of 24 gel pens that I bought from Estrick Stationery Supplies last Thursday.

I need a refund for the pens as they have leaked. The pens were in my bag next to my sports kit, while I was playing football after college. They were still in their plastic case. By the time I got to college and opened my bag there was ink all over the inside and on my sports kit. I have tried to wash them but the ink is permanent and it has ruined my bag and my kit. There is no warning on the packet about this.

When I tried to return the pens to your shop, the assistant was rude and unhelpful. I had my receipt but she told me that I could not have a refund. She told me that it was my fault as I must have left the top off one of the pens.

I would like you to give me a refund for the pens and some money for a new bag and sports kit. I would also like an apology from your assistant.

Make your paragraphs clear: always start a new line for each paragraph. You could leave a whole line clear between each of your paragraphs.

Now try this

Look at the list of ideas for Task D (page 87) on the right.

Structure them into a paragraph plan like the one above. You should:
- group similar ideas together into paragraphs
- number them so that the most important or interesting paragraph is first.

Premier league football team, parks, cafes with outside seating, cricket, some restaurants that deliver, nature walks, football coaching clubs for children, lake with swimming in summer, play areas in park, Chinese and Indian restaurants.

Point-Evidence-Explain

You can structure your paragraphs effectively, by using Point–Evidence–Explain (P.E.E.).
P.E.E. helps you to add detail to your writing and keep it organised.

What is Point-Evidence-Explain?

For every paragraph you should:

- make your point in the first sentence
- provide evidence to support your point.
  This can be more than one sentence if you have
  a lot of details to add.
- explain how the evidence backs up your point.

Evidence

Your evidence should include facts and
details. You can use details from the task
information as evidence, or other details
that you can think of to help to make your
point.

Using P.E.E. to add detail

Look at the paragraph from a student's
sample answer to Task F (page 89) on
the right.

This is the point of the paragraph – it is
about where the money will go.

This is the evidence, or detail, to support
the point. Two sentences are used to give
two facts about equipment in hospitals.

This is the explanation. This explains how
the evidence supports the point.

Sample answer

We'd really like you to provide free
refreshments as we are raising money to
buy equipment for the children's hospital.
Play equipment like art materials and
games can help children recover quickly
from operations. Educational equipment is
also needed to make sure children don't
fall behind at school. New equipment is
very important as some children are in
hospital for a long time.

P.E.E. is a good structure to use when
writing a text to persuade. Evidence
like facts and statistics will help to
make your points more believable and
persuasive.

Getting it right

P.E.E. will help you with both your writing
tasks. Remember to read the task information
carefully to find ideas that you can use as
evidence.

Now try this

Go back to the plan you made on page 53 for Task D on page 87. Use P.E.E. to write up one of the
paragraphs.

Internet discussions

In your writing test, you could be asked to write a blog post or a response to an internet forum discussion.

Blogs and internet forums

Blog posts and **internet discussion messages** tend to be informal and can be shorter than letters or articles. You can write just one paragraph for emails and forum posts, but remember that you will still need to use detail and structure your answer.

> Internet discussions and blogs can be about any topic, including ones you are unfamiliar with. There will be information in the question to help you plan and write your answer.

Writing a forum message

Read Task G on page 90 and look at a student's answer on the right.

✓ The student has included their name and the date at the top left of their answer. This is an important feature of an internet forum message.

✓ The student has written a detailed and structured paragraph using P.E.E.

✓ The student has read the information section clearly and decided to agree with Ben.

> Go to page 53 to revise structuring your writing using Point-Evidence-Explain.

Sample answer

- Audience probably adult
- Internet, so can be informal
- Will need to have strong views
- Agree with Ben

Ellie 2nd February 2016

I agree with you, Ben. Unless we recycle we will lose our countryside to landfill. You should ask the council for a bin, Jake. They provide them free of charge. There is no need to take anything to the tip when you have a bin. The council will take all your recycling and the council do recycle it all. New stuff is made from the recycling. This raises money for the council which covers the cost of transport, fuel and staff. This keeps your council tax down!

point
evidence
explain

Now try this

Read Task G on page 90. Write a plan and write your own message to this internet discussion.

Headings and sub-headings

You could be asked to write an article for your writing task. Articles use headings to get a reader's attention and make the topic clear. They also use sub-headings to divide the writing into smaller sections. This makes the information easier to understand.

Structuring your writing

You should start thinking about structure in your plan. If the task is an article, then headings and sub-headings are a good way to structure your ideas. You should make headings and sub-headings easy to see and useful for the audience by:

- underlining them
- leaving a line before the start of the paragraph
- using a capital letter for the first word
- making it catchy and relevant.

> ### Task 1
>
> **Writing task**
>
> Write an <u>article for the newspaper</u>, giving <u>your views</u> on the subject.
>
> In your article, you should:
>
> - say whether you agree or disagree with the letter
> - say why you agree or disagree
> - give reasons that support your views.
>
> You can add any other interesting information.

Headings

Headings make the topic of an article clear to readers. They should make your article sound interesting so that your audience will want to read on. Here are three different ways to write a heading:

 Make your topic clear.

<u>Mobiles and modern life</u>

This heading makes the topic clear to readers.

 Make a bold statement that shows your views:

<u>Young people need mobiles!</u>

Readers will want to find out why you feel this way.

 Ask a question.

<u>Are mobiles ruining young lives?</u>

This question will make readers think about the topic. They will want to read on to find out how you feel about mobile phones.

Sub-headings

When writing an article, you should use a sub-heading for each of your paragraphs. They should make the topic of the paragraph clear to the reader.

<u>Phones and safety</u>

<u>Phones for information and homework</u>

<u>Phones for friendship</u>

 On your online test, underline headings and sub-headings using this icon 🅰

> Go to page 86 to read the whole of Task 1 above.

Now try this

Read the full task about young people and mobile phones on page 86. Think of some ideas for your article and come up with:

- one heading
- two sub-headings.

Lists and bullet points

Lists and bullet points are a useful way to present information. They allow you to include a lot of detail in your writing and are easy to follow for the reader.

Planning the presentation

Before you start writing, you should decide whether any of the information or details in your plan would work well as a list or as bullet points.

Getting it right

Bullet points and lists are useful ways to present information but you should avoid using them too much in the test. You need to demonstrate all your writing skills in the test, including writing sentences and paragraphs.

Lists

You can use a list to include several items in one sentence. The sentence must still make sense:

The bikes, tennis racquets, footballs and cricket bats can all be inspected at the clubhouse on Saturday, 6 February.

◄ Use commas to separate items in a list. Be careful not to make a list too long: more than four items will make a sentence too complicated long.

Numbered lists

If you are writing to instruct, you can use a numbered list to put your points in order. Each numbered list item should be on a new line.

1. Take a look at our unmissable equipment sale.
2. Contact us to find about more about prices.
3. Arrange a suitable delivery or collection date.
4. Enjoy using your new sports and fitness equipment!

Bullet points

Bullet points are a good way to present information in a report or an advert:

For sale
- 4 ladies' bikes. They all have red and gold frames and one has a shopping basket.

- Tennis racquets. All are in good condition with zipped covers.

◄ Each of these bullet points uses a complete sentence with correct grammar and punctuation. It is a good idea to use complete sentences as much as possible in your test. This will help your writing to make sense.

Now try this

Read the information for this Task J on page 93. Finish the bullet points for the equipment list.

Putting it into practice

You have now revised:

- paragraphs
- Point–Evidence–Explain paragraphs
- headings and sub-headings
- lists and bullet points.

Look at Task D on page 87 and see how a student has used structure to write their answer below.

Sample answer

<u>My local area</u>

<u>Sporting attractions</u>

We are lucky to be the home of two Premier League football teams. Both teams have very big stadiums and you can go on a tour around them. One of the teams has a very successful youth team and runs a terrific weekend coaching club for children. We also have a cricket team that plays every week in the summer. These sporting attractions make the area well worth a visit!

✓ The student has used headings and sub-headings. This makes it clear that the writing is an article.

✗ The heading is not very interesting. Headings need to grab the attention of your audience and make them want to read on.

✓ A sub-heading has been used to tell the reader what the paragraph is about.

✓ A P.E.E. structure has been used here to add detail.

Notice that the student has used one of the bullet points from the task as a sub-heading. You can use the bullet points as sub-headings if they are suitable, or you can think up your own more interesting ones!

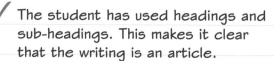

Getting it right

Underline the key words in the task, so that you can easily identify the audience, format and purpose. These three things will help you to decide what **structure** and **language** to use for your writing.

Now try this

Come up with a more interesting heading for the article. Use one of the three heading techniques from page 55. Then write the next paragraph of the article. Remember to:

- stick to one topic or idea in your paragraph
- use a clear P.E.E. structure to add detail
- think of a sub-heading for your paragraph.

Sentences

It is really important that your writing is clear for both the writing tasks. You will need to write complete sentences with correct spelling, grammar and punctuation.

1 Simple sentences

Simple sentences make one point and have **one verb**. A **verb** is an action word.

Every sentence needs a verb and somebody or something to 'do' the verb.

The children swim. ⟵ Verb

The children are doing the swimming.

Ben walks to work. ⟵ Verb

Ben is doing the walking.

2 Adding detail to simple sentences

You can add extra details to simple sentences, to show when, where or how the action is happening.

When

The children swim when it is sunny.

Where

Ben walks to work through the park.

How

Ben walks to work very quickly.

3 Linking simple sentences

To make your writing more interesting, you can put simple sentences together to make longer ones. You can do this by using linking words like:

- **and** to join two points or to add a point.

We went to the zoo **and** we followed the nature trail.

We had chips in the café **and** a coffee.

- **or** to show alternatives.

Visitors can go to the zoo **or** follow the nature trail.

- **but** to add something that disagrees with your point.

We had chips in the café **but** I would rather have had cake.

- **because** to explain your point.

We had chips **because** I was very hungry.

Getting it right

Get into the habit of checking your sentences as you write them:
- Does the sentence have a verb?
- Is it clear who is doing the verb?
- Does the sentence make sense?
- Is the sentence complete? Does it make sense on its own?

Now try this

Read the notes on the right from a student's plan for Task D on page 87. Write a paragraph aiming to use:
- at least **one** simple sentence
- **one** simple sentence with **when, where** or **how** detail added
- at least **one** longer sentence using a **linking word**.

Cinema –
large comfortable seats
café
film club
children's film days

Writing about the present and future

Sentences can be about:

* what is happening now – the present tense
* what has already happened – the past tense
* what will happen in the future – the future tense.

What are verbs?

Verbs (doing words) can change when the tense changes.

Verbs can be:

Action verbs – what somebody or something is doing:

I visit the café every week.

Being verbs – what somebody or something is:

The Nature Reserve café is open on Sundays.

To write about the present or future you need to think about the verb in your sentence.

Present tense verbs

These examples will help you to revise how to use verbs in the present tense:

I visit the café every week.

He visits the nature reserve every Thursday.

They walk through the park every day.

She runs through the park in the morning.

The dog barks when she sees visitors.

We dance on Friday morning.

Janet and Fred jog each evening.

I hate the cold, it makes me shiver.

Add an **s** to the end of a verb for **he, she, it** or **one name.**

Future tense verbs

When you write about the future you can either of the following before the verb:

* **will**

* **going + to.**

The verb does not change when you add **will** before it, and it is the same for **I, you, he, she, it, they** and **we.**

The verb before **going** is not the same for **I, you, he, she, it, they** and **we.**

Add **am** for **I**

Add **is** for **he, she, it** or **one name**

Add **are** for **you, we** or **they**

Notice that if you choose to write 'going', you need to add 'to' before the verb that follows.

I will walk to town tomorrow.

She will visit the nature reserve.

We will do a sponsored walk.

I am going to go to the park.

She is going to do a sponsored swim.

We are going to see the film.

Select the correct verb in each of these sentences:

1 I wakes/wake up every day at 6 a.m. for work.
2 Ben are going/is going to take his mother to the nature reserve next year.
3 They will arrive/arrives at 5 p.m.
4 I love my new alarm clock, it wakes/wake me up very gently.

Writing about the past

To write about things that have already happened, you should use the past tense.

Using the past tense

Like the present and future tenses, to write about the past you need to think about the verb in your sentence.

I watch**ed** a film.
She complain**ed** about the pens.
He decid**ed** to go swimming.

For most verbs, you add **ed** to the end to make them past tense.

He hop**ed** to go to the seaside.
Ben invit**ed** me to his party.

When verbs already end in **e**, just add **d** to the end.

I **tried** hard to get home on time.
Ben **carried** his suitcase to work.
The children **cried** because they were hungry.

Verbs that end in **y** change to **ied**.

I travel**led** a long way.
I stop**ped** eating chips.

For some verbs, you have to double the final consonant (all letters other than vowels) before adding **ed**.

Verbs with their own rules

Some verbs do not follow the rules above. You will need to learn these:

I do > I did	I take > I took	I know > I knew
I have > I had	I go > I went I	I buy > I bought
I see > I saw	make > I made	I bring > I brought
I eat > I ate	I come > I came	I sing > I sang
I get > I got	I sleep > I slept	

Unlike the present tense verbs, the past tense verbs in the table above don't change depending on who or what you are writing about. The verb **to be** is an exception and you are likely to use it often so you should learn it:

I am   > I was
you are   > you were
he/she/it is > he/she/it was
we are   > we were
they are   > they were

Notice that the past tense verbs for **I** and **he/she/it** are the same, and the past tense verbs for **you**, **we** and **they** are the same.

Now try this

Read this student's answer to Task H on page 91. Select the correct verb for the past tense.

We have/had a fantastic day at Estrick Nature Reserve! We followed/follows the nature trail for miles until we were/was tired out. Ben and I see/saw lots of different birds. Ben laughs/laughed as I copyed/copied the bird noises on the way round.

Putting it into practice

You have now revised:

- simple sentences
- adding detail to sentences
- linking sentences
- writing in the past, present and future tenses.

Read Task B on page 85 and look at a student's answer below.

Sample answer

Children played in the park. People walked on the path with their dogs. Mothers used the path with their children. They uses the path to get to school. Slates fall every day from the roof. They could hit any of these people. One of them going to get hurt. Do something. Fix it now. It be a tragedy if a child is hurt.

✗ The wrong tense is being used. The park is being used now, so the present tense is needed.

✗ This paragraph has a lot of short sentences. Some of them could be joined using linking words.

✗ This is the future tense but words are missing.

✗ These are not complete sentences.

✗ The student has not included any features of a letter, such as addresses and names. The information given to you in the task is there to help you. If you are given contact details, you should include them in your answer.

Notice how the incomplete sentences seem rude in a formal letter. Always think about how formal or informal your writing needs to be. Go back to pages 44 and 45 to revise writing style.

Go to page 44 to revise writing a formal letter.

Getting it right

When answering a writing question, remember to:

✓ read the question carefully
✓ underline the audience, purpose and format
✓ make a plan using the task bullet points and information.

Now try this

Correct the student's paragraph above.

1. Make sure all the sentences use the correct tense.
2. Join some of the sentences together using linking words.
3. Change the final three sentences into a more suitable formal style.

When you have completed questions 1–3, check your work.

Full stops and capital letters

In your writing test you will need to use correct punctuation. For both of the writing tasks, you should always:

- start each sentence with a capital letter
- end sentences with a full stop, an exclamation mark or a question mark.

For example:

The nature reserve has many varieties of geese.

Joining sentences

The most common mistake is joining two sentences with a comma instead of using a full stop.

If you are separating two pieces of information or two ideas, you should not use a comma:

✗ The nature reserve has many varieties of geese, visitors are not allowed to feed any of them.

You should use two separate sentences:

✓ The nature reserve has many varieties of geese. Visitors are not allowed to feed any of them.

Or you could join them with a conjunction:

✓ The nature reserve has many varieties of geese but visitors are not allowed to feed any of them.

> Go to page 58 to revise joining sentences.

Questions and exclamations

If you are writing a question, remember to end it with a question mark:

Why don't people recycle glass bottles?

If you want to show excitement or warn of danger, you can end your writing with an exclamation mark:

Our town even has an indoor ski slope!

Be careful when you use exclamation marks. Follow these rules:

- Don't use them too often.
- Never use two or more exclamation marks in a row.

Using capital letters

You should always use a capital letter to start a sentence. You should also use a capital letter at the start of:

- people's names, for example: Mr Malik, Lionel Messi, Mrs Jane Hughes
- names of places, for example: London, Barcelona, Paris
- the month in a date, for example: 20 January 2016.

Names, dates and addresses

Remember to use capital letters correctly for the name, address and date in a letter:

Jane Edwards

Head of the Environment Department

Estrick Council

Main Street

Estrick

ES1 3MS

20 January 2016

Now try this

Rewrite the following sentences. End one with a full stop, one with a question mark and one with an exclamation mark. Make sure you add any missing capital letters.

1 do you want your child hurt by these loose tiles
2 Jane smith suggested I send my complaint to you
3 don't hold a lit firework

Commas and apostrophes

You should use commas and apostrophes in your writing to help readers understand what you mean. Commas are useful for separating information. An apostrophe is used to show that a letter is missing, or to indicate possession.

Apostrophes of possession

Apostrophes are used to show that something or someone belongs to someone or something else. This is called a **possessive** apostrophe:

The boy's hands

Betty's sister

The dog's collar

The school's head teacher

The apostrophe goes between the owner and the 's'. Don't use an apostrophe when you are using an 's' to show a noun is plural.

Banana's/bananas

carrot's/carrots

apple's/apples for sale!

Apostrophes for missing letters

When words are shortened, some letters are missed out. An apostrophe shows where a letter is missing:

cannot > can't

do not > don't

I will   > I'll

let us   > let's

Words that are shortened like this are called **contractions**. They are more informal than the full form. Think carefully about your audience when deciding which to use.

Commas

If you are writing a list, add a comma after each word or phrase:

The nature reserve has birds, small mammals, rare duck breeds and many different varieties of geese.

You do not need a comma between the last two things. You can use 'and' or 'or'.

To prove you are over 18 you will need a driving licence, a birth certificate or a passport.

> Go to page 69 to revise plurals.

Apostrophe warning!

'It's' means it is. The apostrophe shows the letter 'i' is missing.

It's very cold at the nature reserve in the winter.

'Its' means belonging to someone or something. To make sure it is not confused with the contraction for 'it is', there is no apostrophe.

The company opened its fourth shop last year.

This shows that the shop belongs to the company.

Now try this

1 Correct these sentences by adding commas:
   (a) When I visit the reserve I need to take an umbrella a packed lunch and my camera.
   (b) The shop sells postcards guidebooks and wooden toys.
2 Correct these sentences by crossing out the incorrect apostrophe use:
   (a) You must go to Estricks/Estrick's best café!
   (b) Its/It's best to visit the nature reserve in the summer.

Putting it into practice

You have now revised the correct use of:

- full stops, exclamation marks and question marks
- commas
- apostrophes.

Read Task I on page 93 and look at a student's answer below.

Sample answer

i disagree with mary smith! have you thought about the benefits of mobile phones. Many people carry mobile's for directions! Perhaps those young people on their phones were lost!

Its not just young people who need mobile phones while they are out. Older people use mobile's for security. When my grandma is out shopping she worries about falling losing her purse or getting lost. My grandmas mobile makes her feel safe as she can call for help if something frighten's her. Having a mobile with her means she doesnt panic if something go's wrong.

✗ The first two sentences have no capital letter at the start.

✗ Proper names should have capital letters.

✗ This sentence is a question so it needs a question mark at the end.

✗ This paragraph has too many exclamation marks - they lose their impact if more than one is used in a paragraph.

✗ Here, the 's' is used to show that the word is plural so no apostrophe is needed.

You should use correct spelling grammar and punctuation for formal and informal writing.

If you are doing the onscreen test, make sure you remember to capitalise 'I' if you are referring to yourself. Unlike when you are using word processing programs, it won't be done automatically for you.

Getting it right

In your test, remember the following rules for punctuation:

- Start each sentence with a capital letter.
- Use capital letters for proper names.
- Don't over-use exclamation marks.
- End question sentences with a question mark.
- Think about whether words that end in 's' need an apostrophe.
- If you use a contraction, remember to use an apostrophe!

Now try this

There are seven mistakes in the second paragraph of the student's answer above. Rewrite the answer and correct the mistakes.

Spelling

For both your writing tasks you will need to use correct spellings. You can use a dictionary in the test to check spellings, but this will use up a lot of time if you do it too often. This page will give you some helpful spelling tips that will help you to manage without a dictionary.

The i before e rule

As **i** and **e** often appear together, it can be difficult to remember which comes first. Use the **i before e rule** to help you remember:

i before e except after c, but only when it rhymes with bee

believe

rhymes with bee, so i goes before e

receive

rhymes with bee but comes after the c, so e goes before i

eight

doesn't rhyme with bee, so e goes before i

science

comes after c, but it doesn't rhyme with bee, so i goes before e

A few words don't follow the rules: weird, seize, caffeine, species. Practise looking up unfamiliar **ie** or **ei** words in a dictionary before your test.

Ly or ley?

When you add **-ly** to a word ending with **e**, make sure you don't swap the **l** and the **e**. For example:

definite + ly = definitely

bravley　✗　bravely ✓

safley　✗　safely ✓

rudley　✗　rudely ✓

sincerley　✗　sincerely ✓

Make a note of the correct spelling of 'sincerely'. You might have to write a letter for one of your writing tasks. 'Yours sincerely' is one of the ways to end a formal letter.

Words with double letters

Words with double letters can be difficult to spell, as you can't hear the double letters when you say the word. Learn how to spell these words:

address	necessary	tomorrow
different	possible	professional
immediately	occasionally	success
eventually	possible	beginning
opportunity	difficult	recommend
disappoint	disappear	
embarrassing	possession	

Silent letters

Some words have letters that you can't hear when you say the word.

These are some words with silent letters:

when	which	whole
could	knife	autumn
climb	Wednesday	sign
listen	wrong	talk

When you think a word has a double letter, check it in the dictionary. Learn which letters in the word are doubled and which are not.

Now try this

Correct the mistakes in the following sentences:

1　Every paying visitor to the nature reserve recieves a reciept.
2　My mother told me to visit the nature reserve as it has lovley views.
3　Wen I get to the house I cud suprise my mother.

Common spelling errors 1

You will need to use correct spellings for both writing tasks. Be careful to avoid these common spelling errors that are made with words that sound the same but are spelled differently.

 There, their, they're

Their means belonging to them:
Their football boots are muddy.

There is used to explain the position of something:
The football boots were over there.

Or to introduce a sentence:
There is a place for muddy boots outside.

They're is a contraction of **they are**:
They're all tired after the football game.

 We're, wear, where and were

We're is a contraction of **we are**.
We're going to Spain.

Wear is a verb (doing word) that refers to clothing.
You need to wear a uniform at our school.

Were is the past tense of **are**.
They were late getting to the airport.

Where refers to place.
Where are we going?

 Your, you're

Your means belonging to you.

You're is a contraction of **you are**.

✗ ~~Your~~ having the time of ~~you're~~ life

✓ You're having the time of your life.

 To, too, two

To indicates place, direction or position:
I went to Spain.

Two is a number:
Two of us went to Spain last year.

Too means 'also', or a large amount:
I went too far.

 Of, off

The easiest way to remember the difference is by listening to the sound of the word you want to use:

• **of** is pronounced 'ov'
• **off** rhymes with 'cough'

✗ He jumped ~~of~~ the top ~~off~~ the wall.

✓ He jumped off the top of the wall.

 Are, our

Are is a verb (doing word).
We are going to the airport.

Our means belonging to us:
Our football boots are very muddy.

Remember that **a lot** is two words. '**Alot** of people went to Spain' is wrong. It should be: A lot of people went to Spain.

Now try this

Select the correct spelling in these sentences:
1 The boots are two/to/too muddy to go in the car.
2 You are/our lucky to be going on holiday.
3 Students should take their/they're/there books to each lesson.
4 The plane takes off/of in an hour.
5 Your/You're going to love the film.

Common spelling errors 2

You will need to use the correct words for both writing tasks. Be careful to avoid these common errors that are made with words that can sound similar but are spelled differently.

 Would have, could have, should have

Students often use 'would of', 'should of' or 'could of' instead of **would have, could have** or **should have**. For example:

Accidents ~~could of~~ been prevented. We ~~should of~~ fixed the pavement as soon as the cracks appeared.

This student **should have** written: Accidents **could have** been prevented. We **should have** fixed the pavement as soon as the cracks appeared.

 Bought or brought?

Bought and **brought** mean different things.

Bought is the past tense of **buy**:
Ravi bought an umbrella in the shop.

This means Ravi paid money for an umbrella.

Brought is the past tense of **bring**:
Ravi brought an umbrella in her bag.

This means Ravi was carrying an umbrella with her.

 Write or right

Write and **right** mean different things.

Write means to put something in writing, using a pen or pencil:
I need to write a shopping list.

Right is the opposite of **wrong**:
I need to know the right spelling for difficult words.

 Know, no and now

Know means to have knowledge:
I know enough to pass my test.

Now means at the present time:
I now know enough to pass my test.

No is the opposite of **yes**:
'No! That spelling is not correct!'

Now try this

Cross out the incorrect word in these sentences:
1 With my new pen I can right/write in the write/right style.
2 I now/know I should of/should have looked where I was going.
3 I could have/could of bought/brought it cheaper at another shop.
4 The new mobile I brought/bought from Estrick Electronics is faulty.

Common spelling errors 3

You should practise spelling words correctly and build up your vocabulary before your writing test. Some of the most commonly misspelled words are listed below.

Commonly misspelled words

actually	autumn	conclusion	explanation
although	beautiful	decide	February
argument	because	decision	fierce
atmosphere	business	definite	guard
audience	caught	environment	happened
health	modern	persuasion	secondary
heart	nervous	physical	separate
interrupt	performance	preparation	straight
marriage	permanent	queue	survey
meanwhile	persuade	remember	unfortunately

 Learn correct spellings

Get into the habit of looking up new or unfamiliar words in a dictionary before your test. Then practise the correct spelling. You could use the look/cover/write/check method:

1. Look at the word.   2. Write it from memory.
3. Cover the word.   4. Check your spelling.

 Find hidden words

For example, **separate** becomes much easier to spell if you remember there is a 'rat' inside it.

sep-a-**rat**-e

 What you see

Say the word aloud, breaking it into smaller parts. For example, say these words aloud to help you with the correct spelling:

def / in / ite / ly

fri / end

Wed / nes / day

Getting it right

If any of these words are unfamiliar, look them up in a dictionary. That way you will be building up your vocabulary for the test!

Go to page 31 for tips on how to use a dictionary.

Now try this

Spend some time learning these spellings before your test. Test yourself and learn any you get wrong. You could ask somebody to help by testing you.

Plurals

This page explains some useful rules for making words plural. You will need to make sure you use the correct words in both your writing tasks.

Adding s

Most words can be made into plurals by adding **s**:

one test > two tests

a student > lots of students

If a word ends in **ch, sh, x, s** or **ss**, add **es** to make it plural:

one church	two churches
one splash	two splashes
one fox	two foxes
one bus	two buses
one glass	two glasses

Words that end in f or fe

To make a plural when a word ends in **f** or **fe**:

- change the **f** or **fe** to **v**
- add **es**.

loaf	loaves
half	halves
knife	knives
life	lives

Words that end in y

There are different rules for words **ending in y**, depending on the letter **before the y**.

If a word has a **vowel** (a, e, i, o, u) before the **y**, just **add s**:

toys	days	bays
trays	boys	keys

If a word has a **consonant** (any letter that is not a vowel) before the **y**, remove the **y** and add **ies**.

baby	babies
fly	flies
city	cities

Some words don't follow the rules...

Some words use a **different word** as their plural:

woman	women
man	men
foot	feet
tooth	teeth
child	children
person	people
mouse	mice

Some words **don't change** when they are plural:

fish

sheep

deer

It is worth learning the exceptions before your test.

Now try this

Write out the following sentences and cross out the incorrect words.

1 For the partys/parties we need fourteen loafs/loaves of bread.
2 All the torchies/torches need new batteries/batterys.
3 A family ticket covers two adults and up to three childs/children on all buss/buses.
4 Women/womans are just as clever as mans/men.

Checking your work

It is important to leave time at the end of your writing tasks to check your work. Follow these steps to make sure you find any mistakes in your writing.

Careful checking

Always check your work carefully. It is a good idea to check it three times:

- once for spelling
- once for punctuation
- once to check it makes clear sense with no misused, repeated or missing words.

When checking your answers, look for words, punctuation and grammar that you struggle with and focus your attention on these.

Checking for sense

1. When you check for sense, try to read aloud in your head. Imagine you can hear your voice. Does the work still make sense?

2. Remember to leave time to read through your work. Finish both writing tasks before checking, then go back to Task 1.

3. If you come across a sentence that doesn't make sense, read it again. Then think about what you can do to put it right.

Know your strengths

When you are preparing for your test and answering past test questions, you should identify which things you can do well and which things you find difficult. What kinds of mistake do you make:

- spelling mistakes?
- missing or incorrect punctuation?
- missing words?
- using the wrong word?

Alarm bells

Train yourself to hear alarm bells when you come across tricky words, such as:

- their/there/they're
- its/it's
- your/you're.

Stop when you come to any of these words. Double check that you have used the correct spelling for the meaning you intended.

Getting it right

Leaving a brief break before checking your work will help to improve your focus.

Putting it right

If you find a mistake – cross it out. Put ~~one neat line through the mistake~~ and add your correction:

- either by using an arrow (→) to the new words to guide the reader
- or by using an asterisk (*) *to tell the reader to read this bit next.

Now try this

Look at your answer to the task on page 58. Follow the tips above on checking for sense. When you find mistakes, correct them using the **Putting it right** methods above.

Putting it into practice

You have now revised:

• spelling

• checking your work.

Look at Task E on page 88 and read a student's answer below.

Sample answer

Dear Sir

I am writing too you too apply for the job of assistant.

I have previous experiense of working in a shop. I worked part time on Wensdays for too years at Estrick Sandwich Shop. The job involved making sandwichs serving customers and keeping the shop clean and tidy. For this job I needed to be reliable becase I was handling money

You're job is perfect for me becase Im very polit and freindly, I would also like to learn new skilles

My manager at the sandwich shop is mr ben smith and he would be happy to give you a reference. i think he will tell you that I am a reliable and trustworthy employe!

Your sincerley

✗ wrong words used (e.g. 'too' instead of 'to')

✗ spelling mistakes (e.g. 'experiense' instead of 'experience')

✗ missing commas in lists

✗ incorrect plurals (e.g. 'sandwichs' instead of 'sandwiches')

✗ missing apostrophe in 'I'm'

✗ incorrect use of comma

✗ missing full stops at end of sentences

✗ paragraph lacks detail

✗ capital letters missed

Getting it right

In your test you should check your work carefully. It is a good idea to check through your work three times:

✓ once for spelling

✓ once for grammar and punctuation

✓ once to check it makes sense and is clear with no misused, repeated or missing words.

Two of the spelling mistakes here are in words that were used in the information section of the task. If you are unsure of a spelling, check the information section.

Now try this

Rewrite the second paragraph in the student's letter:

• correcting all the mistakes

• adding more detail about why you are suitable for the job.

Putting it into practice (example answer)

Look at a student's answer to Task E on page 88, below. Read the comments to see why this is a good answer.

Sample answer

4 Forest Street
Estrick
ES1 2JY

The Manager
Estrick Mobiles
Main Street
Estrick
EN1 4HH   12th February 2016

Dear Sir/Madam,

I am writing to apply for the job you advertised online for an assistant. I am studying for a BTEC in retail knowledge at Estrick Academy. I would like to work in a shop.

As well as my BTEC I am studying English and Maths. I expect to get C grades in both these subjects in May. My BTEC makes me very suitable for your shop because I am learning about:

- Customer service
- Handling money
- Stock control

In the school holidays I worked at the Estrick Nature Reserve gift shop. When I was there I served customers, handled money and did stock checks. I also worked on a rota so I don't mind those hours.

I'm hard-working, enthusiastic and freindly. I have never been late too school or work. I know a lot about mobile phones. I am very interest in technology.

Yours sincerely,

Maria Ivanovic

What has been done well:

✓ Format is correct for a letter.

✓ The bullet points from the question have clearly been used to plan and structure the answer.

✓ The style is mostly formal, which suits the audience and purpose.

✓ The student has used paragraphs to structure the answer and added detail.

✓ Spelling and punctuation are mostly accurate, with only a few mistakes in the last paragraph.

✓ Grammar is correct. There is only one incorrect tense.

What could be improved:

✗ The style should be formal all the way through. Towards the end, contractions have been used.

✗ Detail has been repeated. It is better to think of something new to say rather than using a detail twice.

Now try this

Use this answer to help you revise.
- Find the spelling errors and correct them.
- Rewrite the third paragraph, using different details about the job in the gift shop.

Putting it into practice (example answer)

Look at a student's answer to Task F on page 89, below. Read the comments to see how this answer could be improved.

Sample answer

To: m.patel@estrickeats.net

sponsored swim

hi maya

my swimming club are doing a sponsored swim thing on 2nd april. Were not paying for the pool can you let us have water bottel's the money will go too the childrens ward at the hospital wich is good as there always needing stuff. lots of people are going to do it four peopel in the pool at once. Bicutes wud be good to! write back

From ben h.

> Go to page 89 to read the task so you can try it yourself.

What has been done well:

✓ The student has used the correct format for an email, with a clear subject line.

✓ The main idea is clear.

✓ The student has used some of the details from the task information.

What could be improved:

✗ The email is not structured in a logical order.

✗ The greeting style is too informal, as the writer does not know the person.

✗ More detail is needed about why the hospital is a good cause.

✗ There are too many spelling errors. Common words should always be spelled correctly.

✗ Some punctuation is missing. Capital letters are needed in places, apostrophes are missing or incorrectly used and some sentences are unclear.

✗ Some sentences are too long and some are incomplete.

Now try this

Rewrite the email making the improvements suggested in the comments above.

Always plan your answer before you write. Planning will help you to cover all the main ideas, add detail and structure the text.

**TEXT A
LETTER FROM ESTRICK ACADEMY**

The letter below was written by an Estrick Academy student to a fashion magazine.

Estrick Academy
14 Main Road
Estrick
ES3 2JJ

The Editor
Fashion Sense Magazine
66 Piccadilly
London
W1

16th May 2016

Dear Editor,

Positive role models in the media

I have decided to write to you to complain about the lack of positive role models in your magazine.

Your magazine is for people between the ages of 13 and 18, who often feel under pressure to 'follow the crowd' so they can fit in with their friends. They look to your magazine for guidance. Your glamorous stories and photographs seem to be all about appearance, even though there is much more to life! You should be inspiring young people to be happy, healthy and ambitious.

At our Academy, we are very disappointed by the lack of positive role models for young people in the media. We carried out a survey and found that over 80 per cent of our students felt under pressure to look or act a certain way because of what they read in magazines like yours. One student said, 'I can't relate to models and film stars. Why is there never anything for sporty people?'.

We think that you should use models with varied appearances, backgrounds, and interests so that there is something for everyone.

You would benefit from making this change by gaining new readers. It would also encourage more parents to buy your magazine for their children. But most of all, think about the positive messages you could be sending out to young people.

Yours sincerely,

Sasha Ali

Estrick Academy

TEXT B
PERSON SPECIFICATION

The person specification below lists the criteria to become a Window Display Assistant.

Person Specification
Window Display Assistant for all Estrick branches of Fashion Fun.

	Criteria	How identified
Qualifications	• Level 2 Functional Skills Maths and English. • At least two GCSEs, including a Design Technology subject such as Textiles, Graphic Design or Resistant Materials.	Certificate
Relevant experience	• No previous experience with window displays is required as training will be provided. • Previous retail experience would be an advantage.	Interview

General requirements	• Must be available at weekends, particularly one Sunday in four.
	
	• Full UK driving licence is essential for this role. You will be responsible for collecting supplies from our central warehouse in our company van.
	
	• Must be willing to undertake a two-year Creative Retail Display course run by Estrick College.
	• Support is given with this qualification through:
	  o on-the-job training and testing
	  o day release to attend college
	  o funding for all course materials.
	• Good mathematical ability is essential.
		Application form
		
		Application form
		
		Interview
		
		Online test at interview

Specialist knowledge and skills	• Creative skills would be an advantage. • Although not essential, the ability to use a sewing machine would be useful.	Interview Application form
Additional factors	• Must be physically fit and healthy, as the job involves some heavy lifting. Reduced member fees for Estrick Gym are offered to all Fashion Fun staff. • Ability to work independently and as part of a team. • Must show a commitment to equal opportunities.	Fitness test Online test at interview Online test at interview

**TEXT C
SAFETY POSTER**

The poster below offers advice on how people can keep their home safe.

SAFETY IN THE HOME

Electrical appliances cause 55% of all fires in the home.

- Keep all electrical appliances away from children.
- Turn off appliances when not in use.
- Do not use electrical appliances in bathrooms or near kitchen sinks – electricity and water do not mix!

Open fires can spark and set furniture and carpets alight.

- Keep all matches and other fire-lighting equipment and flammable materials away from children.
- Do not leave a fire unattended.
- Keep all chemicals and aerosol products away from flames.

Leaving doors and windows unlocked is an invitation to burglars.

- Always lock your doors at night and when you go out.
- Install locks on all your windows and make sure they are closed when you go out.
- Keep valuables out of sight or locked away when not in use.

Many household products contain dangerous chemicals. They can cause serious illness, or even death, if ingested. They can also cause burns .

- Keep all cleaning products away from children.
- Always wear gloves when using strong chemical products.
- Read all labels carefully and make sure you follow the directions.

Safety in the home is very important. Do the following things NOW to make sure your house is safe.

- Share the ideas on this poster with your whole family.
- Check that you are following all the instructions on this poster.
- Contact our helpline to arrange a home safety visit.

Estrick Home Safety Services: Phone: 021 908 7654   Email: estricksafety@estrick.net

**TEXT D
INSTRUCTIONS**

The instructions below tell you how to set up a worm farm.

WONDERFUL WORKING WORM FARM

Ages 10+
Read instructions carefully before using product.

Working toys are a wonderful way to make science fun and easy to understand!

The worm farm has been fully tested, so it is safe to remove the worms and hold them in your hand. There are five worms, so you can share them with your friends. You will be able to watch them grow just like real worms!

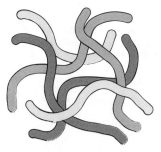

Our website has fascinating facts about how worms grow. You can enter the size of your worms and see if they are bigger than the biggest worm ever recorded! You can also register your worms and have them entered into a prize draw.

Get started now and enjoy your worms!
www.wonderfulwormfarms.com

These instructions are so easy - you can set up your farm all on your own!

1. Fill a large plastic bowl with water. Be careful not to use hot water as this might make your worms grow too quickly.
2. Put the worm mixture into the water. Don't worry if it sinks to the bottom in one big lump – it will soon start to look like worms!
3. Follow the pictures on the enclosed instruction leaflet to put the farm together.
4. After 24 hours (one whole day and night) remove the worms from the water. Put them on pieces of kitchen paper and leave them to dry for an hour.

Put the worms into the farm and watch them grow!

WARNING TO PARENTS OR GUARDIANS

The product is safe for external use only. If swallowed, product could be a throat irritant: you should seek medical help immediately. If heated, vapours can cause headaches.

Call our hotline if any of the parts for this product are faulty:
080 952 3881

**TEXT E
ESTRICK HEALTHY LIVING
CENTRE LEAFLET**

The leaflet below tells you about the Estrick Healthy Living Centre.

The Estrick Healthy Living Centre

Estrick Healthy Living Advice Centre

was set up five years ago by the local council, and we are based in the Town Hall. Our aim is to get local residents following a healthy lifestyle by eating well and exercising more.

What we have achieved so far:

- Set up a free voucher scheme for access to four local gyms
- Provided free health checks for anyone who is worried about their weight
- Provided outdoor gym equipment at the local park
- Set up a healthy living website that offers tips on how to eat healthily.

What we are planning next:

- A series of guided health walks
- A fun day for families
- Evening health sessions at local gyms with a personal trainer.

How you can get involved

We need volunteers to help us spread the word about healthy lifestyles. Volunteers can help in many ways: by raising funds, by working in our local centre, by helping us to plan guided walks. You could also help by delivering leaflets – that way you will also help yourself by keeping fit!

Eat healthily

Start following a healthy lifestyle straight away by trying these menu options:

Vegetarian options

- Tofu hummus with carrot sticks
- Smoky aubergine tagine with lemon
- Pears baked with honey

Low fat option

- Grilled mushrooms
- Thai red duck with sticky sesame rice
- Fruit salad with frozen strawberry yoghurt

Family favourite option

- Spicy lentil and apple soup
- Chicken savoury rice with spicy vegetables
- Fat-free lemon drizzle cake with yoghurt sauce

All these menus have been tried, tested and found to be very tasty!

Recipes for all these menu options are available free of charge on our website: **www.estrickhealthyliving.co.uk**

TEXT F
TRAVEL ARTICLE

The article below, written by John Smith, tells you about Estrick Nature Reserve.

Thinking of a **natural** getaway?

By John Smith – Travel writer

John Smith spent time at one of the three holiday properties now available at Estrick Nature Reserve – an area of outstanding natural beauty within walking distance of stunning coastline walks.

What's there

The nature reserve opened three years ago and has since become the most visited tourist site in the Estrick area, with over 5,000 visitors per month. The money raised from renting the holiday properties helps to improve the habitats of many protected bird species.

During my stay at the reserve I saw reed warblers and red kites. They are only two of the protected species that have come back to the area because of the conservation work done at the reserve. Sharp-eyed birdwatchers can look out for marsh harriers, lapwings and teals, and may even be lucky enough to see the reserve's bean geese.

Visitors using the cottages will be encouraged to take part in the day-to-day running of the reserve by helping to plant trees.

Where to stay

The holiday accommodation sits in five acres of stunning landscaped grounds with its own peaceful lake, tennis courts and indoor swimming pool.

Whilst there, I enjoyed a tasty but healthy lunch snack at the clubhouse. As this sits on the banks of the lake, I didn't need to miss a minute of the breathtaking views.

Woodpecker Cottage sleeps two in a beautiful, king-size four-poster bed. It is a pretty, little property, cosy and peaceful, making it ideal for a romantic break for two.

I was lucky enough to stay in the largest and most luxurious property offered by Estrick Nature Reserve. **Longacre Lodge** is the original hunting lodge which has been lovingly restored and now offers superb accommodation for up to twelve people. It even has a hot tub on the terrace! Like all the holiday properties at the reserve, it comes with a fully-equipped kitchen and a welcome pack is provided on arrival.

If you fancy spending your holiday savings helping to preserve an area of outstanding natural beauty, then have a look at the website:
www.estrickholidaycottages.co.uk.

TEXT G
HAIRDRESSING BLOG

The blog below, written by Fiona, is about hairdressing.

Tales from the Cutting Edge
My hairdressing blog

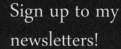

Hi I'm Fiona and I've been a hairdresser for 10 years. I'm writing this blog for hairdressers and people who want to become hairdressers.

Cutting-edge equipment

Posted on 20th April

Sign up to my newsletters!

I was telling a customer today about my very first job in a small salon in a tiny village. Most of our customers were old ladies who wanted a cheap blow dry once a week. The equipment was cheap as the owner needed to keep her prices as low as possible. The problem was, we were constantly replacing the equipment as it was poor quality. At Kids' Cutters we use top-quality hairdryers, scissors and shampoos. Our clients are often as young as three, and wouldn't be able to sit for hours while we worked with blunt scissors!

We also use additive-free shampoos and conditioners. This is really important when working with children as some have allergies, and many are frightened of getting shampoo in their eyes.

Leave a reply:

Archive:
- January
- February
- March
- November
- December

Taking up the clippers!

Posted on 27th April

A customer asked me today how she could become a hairdresser. I told her it is very hard work! You are on your feet all day. The constant hair-washing can make your hands dry. You may not like all of the customers as some may be rude, and some may even refuse to pay. Also, you will always have to work at weekends. But last week I styled the hair of three local girls before their school prom. I had been cutting their hair since they were three years old. Seeing them look so beautiful made the sore feet so worthwhile!

Leave a reply:

**TEXT H
RESTAURANT WORK EXPERIENCE
FACT SHEET**

The fact sheet below tells you about work experience at Estrick Hotel.

Restaurant Work Experience Fact Sheet

You will need to wear:

Your chef's jacket

Sensible shoes or boots

You should bring:

- signed parental consent form
- completed health questionnaire
- £5 to cover food costs
- white shirt and black trousers
- notebook and pen.

For the third year running, Estrick College's catering students have been offered the chance to spend two weeks working at Estrick Hotel. This is a fantastic opportunity to get further knowledge about the catering industry and gain hands-on experience at a top-class hotel kitchen.

Built in 1900, Estrick Hotel has been serving delicious evening meals from its Conservatory Restaurant for over 100 years. Last year, the hotel added a new extension with a café serving light lunches and afternoon teas. The hotel also has a snack bar in the grounds, which provides refreshments for users of its award winning, 18-hole golf course.

Work experience places at Estrick Hotel include:
- Kitchen staff – assisting the chefs with food preparation
- Restaurant staff – waiting on tables
- Bar staff – serving drinks in the bar (only for students who are 18 or over)
- Supplies staff – working with the management team to order ingredients.
- Snack bar staff – making and serving sandwiches, salads and drinks.

You will gain so much from this type of experience.
- Your team-work skills will improve as you will work closely with other hotel staff.
- Your confidence will improve as you will work and be treated like an adult.
- Experience helps your future job prospects. Employers are looking for people with practical experience as well as academic achievements. An employer who sees work experience on your CV will be more likely to hire you.
- As well as these serious benefits, you will also get to eat delicious food every day!

Signing up for work experience is not difficult – you have already taken the first step by picking up this fact sheet.

**TEXT I
ESTRICK VOCATIONAL COLLEGE
WEBSITE**

The website below tells you about Estrick Vocational College.

www.estrickcollege.co.uk/construction

Search

ESTRICK VOCATIONAL COLLEGE

Building trades department

Home | Courses | Careers | Term Dates | Application

Estrick Vocational College was set up 20 years ago to train skilled people to work in local businesses. The Building Trades Department was started 5 years ago and now it runs over 20 courses, including both full-time and part-time.

We have invested in new facilities. Last year, a new building trades centre was added. It has workshops for bricklaying, joinery, painting and decorating, plastering, plumbing and electrical installation work. We aim to prepare students for life on a real construction site. We also built new classrooms and an assembly hall. This year, we have spent over £2 million on a state-of-the-art computer suite. These improvements will support students with the academic side of their college courses.

Why choose Estrick College?

- All our courses are industry-approved.
- Our workshop facilities meet all health and safety standards.
- We provide all materials and tools.
- Our staff are fully qualified and friendly.
- We offer support with English, Maths and IT skills.
- Over 75% of students find a construction industry job within one month of leaving college.

 Why do employers choose Estrick College students?

Our students are fully prepared for the workplace. All students have experience of working on real construction sites. For the past 3 years all our students have left the college with industry-approved qualifications in their particular trade subject.

 What are the entry requirements?

You must have a grade C in Maths and English. Applicants with a grade D in these subjects may be considered if they show excellent potential in their chosen trade.

 How do I find out more about the courses?

Click on the appropriate tabs to find out more information about all our courses.

 How do I apply?

Complete our application form and send it to us. Our address can be found on our home page.

**TEXT J
CHARITY REPORT**

The charity report below tells you about the fundraising for Estrick Animal Centre.

REPORT INTO CHARITY SPENDING

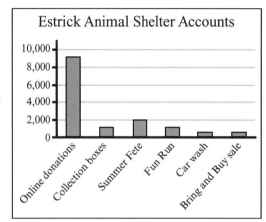

To: Charity Fundraisers

From: Chairman, Estrick Animal Shelter Committee

Subject: Fundraising (October 2014 – October 2016)

Date: 18th December 2016

Dear Charity Fundraisers,

I am writing to give you the details of our fundraising, from October 2014 to October 2015. Together, we raised a lot of money for animal welfare projects in 2015, but we know that you can help us raise even more money in 2016.

The chart on the right shows the amount of money raised by each fundraising activity. Read on to find out more.

Donations

The animal shelter still relies mainly on donations from the public. Donations have increased by 50 per cent and have now reached over £10,000 since we added the new online donation feature on our website. This means that people do not have to leave home to donate. It has also encouraged younger people to donate. Online donations are higher than those received in collecting boxes.

Charity Events

The Summer Fete was extremely successful and raised over £2,000. We were blessed with good weather and the local Brownies did a fantastic job setting up all the stalls. Cake sales accounted for half of the money raised at the fete. We were lucky to have held the fete while *The Great British Bake Off* was on television! Next year, we could double the amount raised at this event if we can encourage more volunteers to run stalls.

For the second year running, our Fun Run was held during a torrential downpour. This did not dampen the spirits of the runners, who managed to raise a total of £1,000 in sponsorship! This event is growing in popularity and we will need more volunteers along the route next year.

Unfortunately, the car wash event was poorly attended. The Scouts did an excellent job but we will need to ask for more volunteers next year to advertise this event. The same goes for the Bring and Buy sale, where we need people to ask local companies to donate more exciting products.

TASK A

In the test this would be a Task 1 writing task so it would be worth 15 marks.

INFORMATION

You bought a set of pens from Estrick Stationary Supplies. The first time you put the pens into your bag one of them leaked onto your sports kit.

When you returned to the shop the assistant was rude and refused to refund your money for the pens.
You decide to write to the manager.

WRITING TASK

Write a letter of complaint to Asha Sharif Manager of Estrick Stationary Supplies Ltd, 8 High Street, Estrick, EW12 5HH.

You may wish to include:

- the product details, including where and when you bought the pens
- what went wrong with the pens
- how you have been treated by the shop assistant
- what you want to happen now.

Remember to set out your letter using the correct layout.

TASK B

> In the test this would be a Task 1 writing task so it would be worth 15 marks.

INFORMATION

Walking through your local park, you notice the poor condition of the roof of the community centre. A large number of slates are loose and you are worried that they will fall onto the path and hit somebody. This could be very dangerous because there are always a lot of young children in the park.

WRITING TASK

Write a letter to Malin Edwards, Head of the Environmental Department, Estrick Council, Main Street, Estrick, ES1 3MS, to inform her of this problem.

In your letter you should:

- introduce yourself
- describe the problem
- explain who could be at risk
- say what you want the council to do about it.

Remember to set out your letter using the correct layout.

TASK C

In the test this would be a Task 1 writing task so it would be worth 15 marks.

INFORMATION

You notice this letter in your local newspaper.

Estrick News

20 August 2016

Letter of the week

Dear Editor,

I have noticed that most young people seem to spend all day on their mobile phones. Why? What is so interesting that they need to ignore what is happening around them?

Last week a young man bumped into me on the pavement because he didn't look up from his phone. I could have fallen and hurt myself badly.

I think young people should only be allowed to use their phones for two hours a day. They should also not be allowed to use their phones in public.

Yours sincerely,

Anonymous

WRITING TASK

Write an article for the newspaper, giving your views on the subject of this letter.

In your article, you should:

• state whether you agree or disagree with the letter
• say why you agree or disagree
• give reasons that support your views.

You can add any other interesting information.

TASK D

INFORMATION

A group of international exchange students is coming to your town to spend a month learning English. They are keen to visit all the attractions in your local area.

You decide to let them know more about your local area by writing about what there is to do and see. You have heard they like sport, eating out and walking.

WRITING TASK

Write an article for your school/college/workplace newspaper, describing the kinds of things the international exchange students can do in your area.

In your article you may want to include details of:

- local attractions
- leisure facilities
- shopping and local markets.

You could add any other useful information.

TASK E

In the test this would be a Task 1 writing task so it would be worth 15 marks.

INFORMATION

You see the following advertisement online and decide to apply.

Estrick Mobiles

Assistant required

We are looking for an assistant to work in our mobile phone shop on Estrick Main Street.

Applicants must be hard-working, enthusiastic and friendly. Previous experience in a shop environment would be an advantage, but is not essential.

We are open until 8 p.m. and open on Sundays, so applicants must be prepared to work different hours each week on a rota.

If you are interested in applying, please send a letter to the manager:

**Estrick Mobiles,
Main Street,
Estrick,
EN1 4HH.**

WRITING TASK

Write a letter to the manager of Estrick Mobiles, applying for the job.

In your letter you should:

- explain why you are writing
- explain why you want the job and why you would be suitable
- give details of your experience.

You may add any other ideas you have.

Remember to set out your letter using the correct layout.

TASK F

In the test this would be a Task 2 writing task so it would be worth 10 marks.

INFORMATION
You are a member of a group about to take part in a sponsored swim to raise money for a local charity. You receive an email from the group's leader asking you to arrange refreshments.

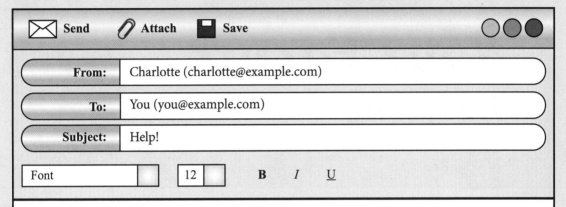

From:	Charlotte (charlotte@example.com)
To:	You (you@example.com)
Subject:	Help!

Font   12   **B**   *I*   U̲

Hi,

Our sponsored swim has now been set for Sunday 2nd April. We are hoping that about 30 people will take part.

Estrick Leisure Centre have agreed to let us use their pool. However, we will need to bring our own refreshments.

Please can you get in touch with Maya Patel at Estrick Eats to persuade her to provide tea, coffee and water? I'm sure the swimmers would quite like some biscuits too!

Don't forget to mention that the swim is to raise money to decorate the children's ward at the local hospital.

Her email address is m.patel@estrickeats.net.

I know you will do a great job for us!

Charlotte

WRITING TASK
Write an email to Maya Patel, the owner of Estrick Eats.
In your email you should:
• give details of the sponsored swim
• persuade him to provide the refreshments.

TASK G

In the test this would be a Task 2 writing task so it would be worth 10 marks.

INFORMATION

You have seen the following messages on an Internet discussion about recycling.

Your Forum
Recycling

 Ben   20th January 2016

I can't believe people don't bother to recycle. The council provide a bin for this purpose! If you don't have a bin you can take stuff to the tip. Surely everyone knows that recycling is important for the environment. We are wasting all the planet's natural resources by throwing everything away. People need to start being responsible about their rubbish.

 Eva   30th January 2016

Ben, you might be lucky enough to have a recycling bin – I don't have one! I don't have time to take stuff to the local tip as it takes ages to sort everything into different bins. To be honest, I don't think it gets recycled anyway. Also, I bet transporting rubbish to the tip uses loads of fuel, which is also damaging to the environment!

Response

WRITING TASK

Write your own message in response to this Internet discussion, giving your detailed views on the subject of recycling.

Use sentences and write in Standard English.

TASK H

In the test this would be a Task 2 writing task so it would be worth 10 marks.

INFORMATION
You and your friends recently had a day out at Estrick Nature Reserve.

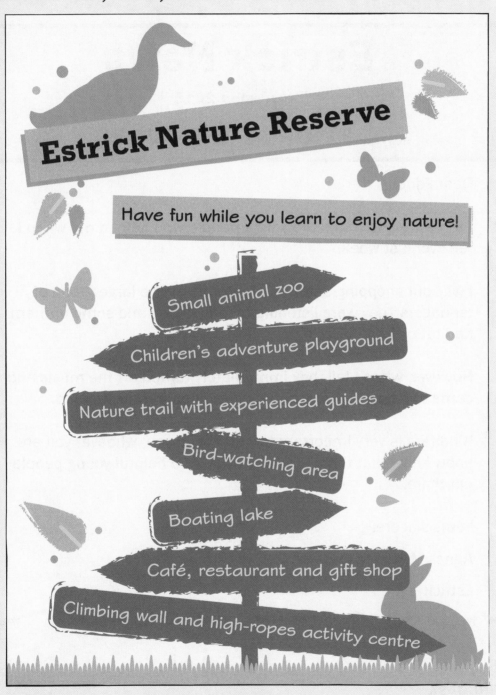

Estrick Nature Reserve

Have fun while you learn to enjoy nature!

- Small animal zoo
- Children's adventure playground
- Nature trail with experienced guides
- Bird-watching area
- Boating lake
- Café, restaurant and gift shop
- Climbing wall and high-ropes activity centre

WRITING TASK
Write an email to your friend, telling them about your day out at Estrick Nature Reserve.
In your email you may:
- explain what you liked **or** disliked about Estrick Nature Reserve
- say why you think it would be a good **or** a bad place to visit.

TASK I

In the test this would be a Task 2 writing task so it would be worth 10 marks.

INFORMATION
You noticed this letter in your local newspaper.

Estrick News

1 August 2015

Letter of the week

Dear Editor,

I am writing to thank the young people who helped me when I fell over last week.

I was out shopping and was frightened by the large group of teenagers. They were listening to loud music and some of them had tattoos and piercings.

However, when I fell they immediately turned off the music and came to help me. None of the older people even stopped.

Kindness is very important to me. Thank you, whoever you are. I now know that there are some kind and helpful young people out there.

Yours sincerely,

Agnes Mencher

Estrick

WRITING TASK
Write an email to editor@estricknews.org giving your views on the letter.
You may wish to include:
- where and when you saw the letter
- your views on the content of the letter.

TASK J

> In the test this would be a Task 2 writing task so it would be worth 10 marks.

INFORMATION
You are a member of a local football club and receive this email from the club secretary.

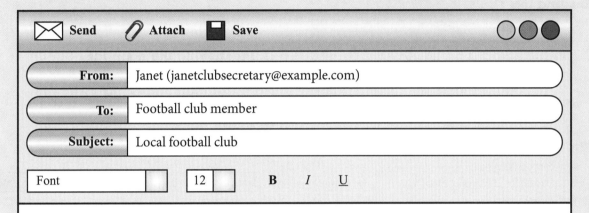

✉ Send   📎 Attach   💾 Save   ○○●

From:	Janet (janetclubsecretary@example.com)
To:	Football club member
Subject:	Local football club

Font ▢   12 ▢   **B**   *I*   <u>U</u>

Hi!

Thanks for helping to tidy up the clubhouse last weekend! We found loads of old sports equipment that we can now sell, including:

- 4 bikes – all need new chains
- 6 tennis racquets – good condition
- 4 footballs – good condition
- 10 cricket bats – useable, but not great condition.

We are short of money this year and need to redecorate the clubhouse. Selling this equipment would be a great way to raise money quickly!

Can you write an advert for this equipment, please? I can then put it into the local newspaper. Give my phone number as I don't mind taking the calls.

Thanks! I know you are good at this type of writing!

Janet

WRITING TASK
Write the text for the advert, describing the equipment.

In your advert you may wish to include:
- a detailed description of the items
- your reason for selling
- the price you are asking
- how buyers can contact the club.

Answers

READING

1. Your reading and writing test

1 Two

2 Check your work

2. Reading test skills

1 The theme or topic of the texts

2 Yes

3 Answers could include any two of the following:
- underlining
- bold font
- bullet points
- numbered lists

3. Multiple choice questions

1 Four or five options

2 Put a line through the wrong cross and then mark new answer with a cross.

4. Short response questions

1 Answers could include:
- Do not write in full sentences
- Give only one answer per line
- Read the question carefully and identify key words

2 All information for the answers can be found in the texts.

5. Reading the question

1 2 answers

2 You will not get marks for information that is not in the texts.

6. Skimming for ideas and details

1 For example:
- Bullet points
- The first sentence of each paragraph

2 So you can focus your reading on the details that will help you answer the question

7. Underlining

How you can get involved

We need volunteers to <u>help us spread the word about healthy lifestyles</u>. Volunteers can help in many ways: by <u>raising funds, by working in our local centre, by testing the recipes on our website</u> or <u>by helping us to plan guided walks</u>. You could <u>help by delivering leaflets</u> – that way you will also help yourself by keeping fit!

8. Online tools 1

1 Click the Time button.

2 Change the colour or font size.

9. Online tools 2

1
- Underlining questions
- Underlining words/phrases
- Making a plan for answering writing tasks

2 Flag it and go back to it later.

10. Putting it into practice

A. reed warblers

D. red kites

11. Understanding the main idea

1 By looking at the features.

2 Answers could include any two of: inform, instruct, describe or persuade

12. Identifying the main idea

1 Types of work experience at the Estrick Hotel

2 It is made clear by the heading and the first sentences.

13. Types of text

1 For example: search bar and links = website

2 For example:

letters	articles	websites	blogs
Addresses	Headings/titles	Navigation bar	Dated sections
'Dear'	Sub-headings	Search box	Links
date	paragraphs	Links	Comments
Paragraphs	Pictures	Hyperlinks	
Sign-off = Yours sincerely	columns		

14. More types of text

1 Images

2 Answers could include any two of the following:
- facts
- statistics
- charts
- bullet points
- tables

3 To make information easier to find.

15. Texts that inform

1 (a) Answers could include any two from the following:
- Meals have been served for over 100 years.
- Last year the hotel added a new café serving light lunches and afternoon teas.
- The hotel has a snack bar.
- The hotel has a golf course.

(b) For instance: 'assisting' and 'management team'

2 What work experience is available at the hotel and how students can gain it.

16. Texts that instruct

1 Answers could include any two of the following:
- Command verbs – 'keep'
- Concise sentences – 'Use a fireguard at all times.'
- Clear language – 'near kitchen sinks'

2 For example – it uses command verbs so that readers do exactly the right thing and it uses clear language so that readers can easily understand what to do.

3 Keep, turn off, do, use, leave, wear, install, lock, install, read, check, share, contact.

17. Texts that describe

1 Answers could include:
- Bold fonts
- Outstanding natural beauty
- Stunning coastline walks
- Nutcracker Cottage is the perfect place to get away from the hassle of modern life.

2 Answers could include:
- Luxurious
- Superb
- Beautiful
- Pretty
- Fully-equipped

18. Texts that persuade

Facts could include:
- College was set up 20 years ago
- Computer suite cost £2 million
- Over 75% of students find a construction industry job

Details could include: details about facilities, details about what college provides.

19. Putting it into practice

Underlining should identify key information in first lines of paragraphs,

20. Understanding presentation

Answers could include:
- Bullet points
- Paragraphs
- Columns
- Images

21. Paragraphs, columns and fonts

1 Answers could include: different font, size, colour, or bold

2 Answers could include: Text A, Text D, Text E, Texts G–J

22. Titles, headings and lists

1 So that you can fully understand the main ideas

2 Answers could include: Texts B–E, Text H, Text J

23. Tables and charts

For example:

1 Tables, graphs and charts make facts easy to find.

2 Tables, graphs and charts mean you can find information quickly.

24. Images, colours and graphics

Texts A–J all use a combination of images, colours and graphics.

25. Putting it into practice

Answers could include any two of the following: different fonts, graphics or images, bullet points, title or heading.

26. Understanding detail

- How can you find out more about the college courses?
- Give **one** reason why employers choose students from the college.
- The Building Trades Department of the College was set up:
- Which **one** of the statements about the college is true?

27. Reading for detail

A and E.

28. Careful reading

Because some of the statements in the question may not be written as they are in the text.

29. Tricky questions

Answers should include as many of the following:
- read questions carefully
- only use information from the text
- avoid reading too much into the text
- go back at the end and check your answer

30. Vocabulary

As the word appears in a sentence about products causing death or serious illness, 'ingested' probably means swallowed.

31. Using a dictionary

Vapours – 'a substance diffused or suspended in the air, especially one normally liquid or solid', so it means the fumes from the product.

32. Putting it into practice

1 D, **2** EITHER enter the size of your worms OR register your worms **3** call the hotline

33. Using information

Welcome pack on arrival

34. Responding to a text

Answers should include any three of the following:
- Running a stall at the Summer Fete
- Volunteering at the Fun Run
- Volunteering at the car wash event
- Volunteering at the Bring and Buy

35. Putting it into practice

1 Answers could include any two of the following:
- you are on your feet all day
- rude customers
- hands gets dry
- customers who refuse to pay
- working at weekends

2 Answers could include any two of the following:
- top-quality hairdryers, scissors and shampoos
- professional, caring attitude
- additive-free shampoos and conditioners
- awareness of children's allergies and fears

36. Avoiding common mistakes

1 Your friend is thinking about buying a worm farm for his son, but worries that his son will find it boring.

Using the information in Text D, give two points to convince your friend that his son will not find the worm farm boring.
- you can share the worms with friends
  Second answer could be one of the following:
- you can watch them grow like real worms
- the worms keep growing for a whole year
- you can register them on the website and have them entered in a prize draw

2 According to Text D, what should you do if the product is faulty?

Call the hotline.

37. Checking your work

1 Identify <u>two</u> activities from Text D that students will be able to do <u>during their work experience</u>.

Answers could include any two of the following:
- assisting the chefs with food preparation
- waiting on tables
- if over 18, serving drinks in the bar
- making and serving food and drinks at the snack bar
- helping to order supplies

2 According to Text D, <u>how</u> do you <u>register your interest in work experience</u> at the hotel?

By logging on to the school website

WRITING

38. Writing test skills

1 To make sure the reader has all necessary or important information.

2 Answers could include any two of the following:
- paragraphs
- numbered lists
- headings
- sub-headings

39. Writing test tasks

Notes could include:
- letter to somebody not known so formal language,
- detail needed to develop points,
- use information about why pens faulty, how assistant behaved and want I want.

40. Putting it into practice

Timing should be:

Task C: planning 5 minutes, writing 18 minutes, checking 2 minutes

Task H: planning 5 minutes, writing 13 minutes, checking 2 minutes

Task H planning notes, for example:

Informal language, paragraphs, no 'Dear', use bullet points and add details to make it sound fun

41. Understanding audience

For example:

Audience – college students, will probably need the library for studying; Local attractions - would enjoy cinema, leisure centre, castle as it is historic, might like art gallery if art student; Shops – large shopping mall with fashion and sports shops.

42. Letters and articles

1 Answers could include any three of the following: Your address, recipient's address, date, greeting, sign-off

2 Capital letters

43. Emails and online discussions

1 When you know the person you are writing to.

2 For example

Format	Layout
Letters	Date, addresses, Yours sincerely
Articles	Headings, sub-headings, use capital letters
Emails	Heading line, informal ending
Internet forum discussions	Date, name

44. Formal writing

For example:

The path goes straight through the park and is used by children on their way to school. Young children also use the path to get into the community centre.

45. Informal writing

For example:

You can just chill out in the café and watch all the rare birds on the lake. You'll be able to show off in the boats but don't scare the geese!

46. Putting it into practice

For example, if audience is workplace:

<u>Local attractions</u>

Estrick has a lot of interesting historic places to visit. The castle has walls that are five foot thick. Children will enjoy walking all the way around the walls. There are history talks at the weekends which can be a real help with your children's homework. They give discount for families and students.

<u>Leisure facilities</u>

Estrick has a superb new leisure complex right in the middle of town. It has an Olympic-size swimming pool and children will love going down the slides. The pool is not just for serious swimming. It has a wave machine which makes it fun for the whole family. There is also a café that sells children's meals. It is worth visiting the town just to spend a whole day here.

47. Understanding planning

Plans could include:
- My name
- Slates loose
- Children who use park, children walking to school, joggers, people using the community centre
- Replace loose slates immediately

48. Using detail

Plans could include:
- Saw letter in paper this week
- Agree - not all teenagers are frightening
- People shouldn't judge on appearance like tattoos and piercings
- Loud music doesn't mean trouble
- Most young people are kind

49. Ideas

Additional details could include:

- Slates look jagged and sharp
- Loose slates are very close to the edge of the roof
- Holes in the roof where tiles have come loose

50. Putting it in order

1 My name

2 Slates loose; Slates look jagged and sharp

3 Children who use park, children walking to school, joggers, people using the community centre

4 Replace loose slates immediately

51. Putting it into practice

Effective plans will be detailed, cover both bullet points and be in logical order, for example:

1 Saw letter this week in Estrick News, needed to give own view about teenagers

2 Shouldn't judge on appearance - tattoos and piercings aren't bad, just fashion

3 Loud music means enjoying themselves not aggressive

4 Most teenagers are kind and helpful

4 Not surprised that teenager was kind, all my friends are [give example]

52. Using paragraphs

Answers should group ideas into three paragraphs. For example:

Sports –

1 Premier League football team

2 football coaching clubs for children

3 cricket

Outdoor attractions –

1 parks

2 play areas in park

3 nature walks

4 lake with swimming in summer

Eating –

1 cafes with outside seating

2 Chinese and Indian restaurants

3 some restaurants that deliver

53. Point-Evidence-Explain

For example:

[Point] My local area has many sporting attractions for all the family. [Evidence] We have a Premier League football club that won the league last year. [Explanation] The club runs football coaching clubs for children.

54. Internet discussions

Effective answers should:

- use an informal style
- respond to the comments made by Ben and Jake
- include additional details and opinions
- use correct spelling, punctuation and grammar

55. Headings and subheadings

For example:

Heading [using a question]: Are mobiles the new youth addiction?

Sub-headings: Using a mobile for safety; Using a mobile for information.

56. Lists and bullet points

Bullets should contain at least one full sentence.

For example:

- Footballs. All in good condition.
- Cricket bats. The bats are a bit battered and scratched but will be good for training games.

57. Putting it into practice

Effective answers should use one of the three heading techniques from page 56, and have a one-idea/topic paragraph that uses a clear P.E.E. structure.

58. Sentences

For example:

Our town has a multiplex cinema.[simple sentence] It has large comfortable seats that recline for extra comfort. [simple sentence with detail added about *how*]The café serves lovely meals before every evening film. [simple sentence with detail about *when*]. The cinema runs a film club and film days for children. [simple sentences joined with *and*]

59. Writing about the present and future

1 I wakes/wake up every day at 6am for work.

2 Ben are going/is going to take his mother to the nature reserve next year.

3 They will arrive/arrives at 5pm.

4 I love my new alarm clock, it wakes/wake me up very gently.

60. Writing about the past

We have/had a fantastic day at Estrick Nature Reserve! We followed/follows the nature trail for miles until we were/was tired out. Ben and I see/saw lots of different birds. Ben laughs/laughed as I copyed/copied the bird noises on the way round.

61. Putting it into practice

For example: Children play in the park and people walk on the path with their dogs. Mothers use the path with their children and children use the path to get to school. Slates fall every day from the roof because they are loose. They could hit any of these people and one of them is going to get hurt.

I think the Council should do something to fix the slates soon because it will be a tragedy if a child is hurt.

62. Full stops and capital letters

1 Do you want your child hurt by these loose tiles?

2 Jane Smith suggested I send my complaint to you.

3 Don't hold a lit firework!

Answers

63. Commas and apostrophes

1 **(a)** When I visit the reserve I need to take an umbrella, a packed lunch and my camera.

  (b) The shop sells postcards, guide books and wooden toys.

2 **(a)** You must go to ~~Estricks~~/Estrick's best café!

  (b) ~~Its~~/it's best to visit the nature reserve in the summer.

64. Putting it into practice

It's not just young people who need mobile phones while they are out. Older people use mobiles for security. When my grandma is out shopping she worries about falling, losing her purse or getting lost. My grandma's mobile makes her feel safe as she can call for help if something frightens her. Having a mobile with her means she doesn't panic if something goes wrong.

65. Spelling

1 Every paying visitor to the nature reserve receives a receipt.

2 My mother told me to visit the nature reserve as it has lovely views.

3 When I get to the house I could surprise my mother.

66. Common spelling errors 1

1 The boots are ~~two/to~~/too muddy to go in the car.

2 You are/~~our~~ lucky to be going on holiday.

3 Students should take their/~~they're/there~~ books to each lesson.

4 The plane takes off/~~of~~ in an hour.

5 ~~Your~~/you're going to love the film.

67. Common spelling errors 2

1 With my new pen I can ~~right~~/write in the ~~write~~/right style.

2 I ~~now~~/know I ~~should of~~/should have looked where I was going.

3 I could have/~~could of~~ bought/~~brought~~ it cheaper at another shop.

4 The new mobile I ~~brought~~/bought from Estrick Electronics is faulty.

68. Common spelling errors 3

Have you practised the tricky spellings?

69. Plurals

1 For the ~~partys~~/parties we need fourteen ~~loafs~~/loaves of bread.

2 All the ~~torchies~~/torches need new batteries/~~batterys~~.

3 A family ticket covers two adults and up to three ~~childs~~/children on all ~~buss~~/buses.

4 Women/~~womans~~ are just as clever as ~~mans~~/men.

70. Checking your work

Have you checked the plans and answers that you have written already for mistakes?

71. Putting it into practice

I have previous experience of working in a shop. I worked part time on Wednesdays for two years at Estrick Sandwich Shop. The job involved making sandwiches, serving customers and keeping the shop clean and tidy. For this job I needed to be reliable because I was handling money.

72. Putting it into practice (example answer)

Effective answers should have 'friendly' and 'interest' corrected in the final paragraph. The third paragraph should use different detail about the gift shop job.

73. Putting it into practice (example answer)

Effective answers should:
- use a formal style
- be logically structured
- correct sentences, with sentences joined where appropriate
- include plenty of detail
- a minimum of spelling and punctuation errors.

Published by Pearson Education Limited, 80 Strand, London, WC2R 0RL.

www.pearsonschoolsandfecolleges.co.uk

Copies of official specifications for all Edexcel qualifications may be found on the website: www.edexcel.com

Text © Pearson Education Limited 2016
Edited by Jane Anson
Typeset by Jouve India Private Limited
Produced by Elektra Media
Original illustrations © Pearson Education Limited 2016
Illustrated by Elektra Media
Cover illustration by Miriam Sturdee

The right of Julie Hughes to be identified as author of this work has been asserted by her in accordance with the Copyright, Designs and Patents Act 1988.

First published 2016

19 18 17 16
10 9 8 7 6 5 4 3 2 1

British Library Cataloguing in Publication Data
A catalogue record for this book is available from the British Library

ISBN 978 1 292 14580 8

Printed in Italy by Lego S.p.A.

Acknowledgements
The author and publisher would like to thank the following individuals and organisations for permission to reproduce photographs:

(Key: b-bottom; c-centre; l-left; r-right; t-top)

Fotolia.com: koss13 20b, 79b; **Shutterstock.com:** Donna Ellen Coleman 35, 81b, keko64 14, 20t, 24, 79t, New Punisher 50, 86, racorn 2, 81t

A note from the publisher
In order to ensure that this resource offers high-quality support for the associated Pearson qualification, it has been through a review process by the awarding body. This process confirms that this resource fully covers the teaching and learning content of the specification or part of a specification at which it is aimed. It also confirms that it demonstrates an appropriate balance between the development of subject skills, knowledge and understanding, in addition to preparation for assessment.

Endorsement does not cover any guidance on assessment activities or processes (e.g. practice questions or advice on how to answer assessment questions), included in the resource nor does it prescribe any particular approach to the teaching or delivery of a related course.

While the publishers have made every attempt to ensure that advice on the qualification and its assessment is accurate, the official specification and associated assessment guidance materials are the only authoritative source of information and should always be referred to for definitive guidance.

Pearson examiners have not contributed to any sections in this resource relevant to examination papers for which they have responsibility.

Examiners will not use endorsed resources as a source of material for any assessment set by Pearson.
Endorsement of a resource does not mean that the resource is required to achieve this Pearson qualification, nor does it mean that it is the only suitable material available to support the qualification, and any resource lists produced by the awarding body shall include this and other appropriate resources.